PENGUIN BOOKS

THE GHOST MAP

Steven Johnson is the author of the acclaimed books *Everything Bad is Good for You* (described as a 'must read' by Mark Thompson, head of the BBC), *Mind Wide Open*, *Emergence* and *Interface Culture*. His writing has appeared in the *Guardian*, the *New Yorker*, *Nation* and *Harper's*, as well as the op-ed pages of *The New York Times* and the *Wall Street Journal*. He is a Distinguished Writer in Residence at NYU's School of Journalism, and a Contributing Editor to *Wired*. He is also the co-creator of several influential websites: FEED, Plastic and Outside in. He has degrees in Semiotics and English Literature from Brown and Columbia Universities. He lives in Brooklyn with his wife and three sons.

Steven Johnson hosts a web log at www.stevenberlinjohnson.com

ALSO BY STEVEN JOHNSON

INTERFACE CULTURE:
How New Technology Transforms the Way
We Create and Communicate

EMERGENCE:
The Connected Lives of Ants, Brains,
Cities, and Software

MIND WIDE OPEN:
Your Brain and the Neuroscience
of Everyday Life

EVERYTHING BAD IS GOOD FOR YOU:
How Today's Popular Culture Is Actually Making Us Smarter

The GHOST MAP

A street, an epidemic and the hidden power of urban networks

STEVEN JOHNSON

PENGUIN BOOKS

PENGUIN BOOKS

Published by the Penguin Group
Penguin Books Ltd, 80 Strand, London WC2R 0RL, England
Penguin Group (USA) Inc., 375 Hudson Street, New York, New York 10014, USA
Penguin Group (Canada), 90 Eglinton Avenue East, Suite 700, Toronto, Ontario, Canada M4P 2Y3
(a division of Pearson Penguin Canada Inc.)
Penguin Ireland, 25 St Stephen's Green, Dublin 2, Ireland (a division of Penguin Books Ltd)
Penguin Group (Australia), 250 Camberwell Road,
Camberwell, Victoria 3124, Australia (a division of Pearson Australia Group Pty Ltd)
Penguin Books India Pvt Ltd, 11 Community Centre,
Panchsheel Park, New Delhi – 110 017, India
Penguin Group (NZ), 67 Apollo Drive, Rosedale, North Shore 0632, New Zealand
(a division of Pearson New Zealand Ltd)
Penguin Books (South Africa) (Pty) Ltd, 24 Sturdee Avenue,
Rosebank, Johannesburg 2196, South Africa

Penguin Books Ltd, Registered Offices: 80 Strand, London WC2R 0RL, England

www.penguin.com

First published in the United States of America by Riverhead Books 2006
First published in Great Britain by Allen Lane 2006
Published in Penguin Books 2008
1

The passage from Walter Benjamin's "Theses on the Philosophy of History"
is from *Illuminations,* translated by Harry Zohn.

A list of illustration credits can be found on page 300.

Printed in Great Britain by Clays Ltd, St Ives plc

978-0-141-02936-8

For the women in my life:

My mother and sisters, for their amazing work
on the front lines of public health

Alexa, for the gift of Henry Whitehead

and Mame, for introducing me to London so many years ago . . .

CONTENTS

A Klee painting named "Angelus Novus" shows an angel looking as though he is about to move away from something he is fixedly contemplating. His eyes are staring, his mouth is open, his wings are spread. This is how one pictures the angel of history. His face is turned toward the past. Where we perceive a chain of events, he sees one single catastrophe which keeps piling wreckage and hurls it in front of his feet. The angel would like to stay, awaken the dead, and make whole what has been smashed. But a storm is blowing in from Paradise; it has got caught in his wings with such a violence that the angel can no longer close them. The storm irresistibly propels him into the future to which his back is turned, while the pile of debris before him grows skyward. This storm is what we call progress.

—Walter Benjamin, "Theses on the Philosophy of History"

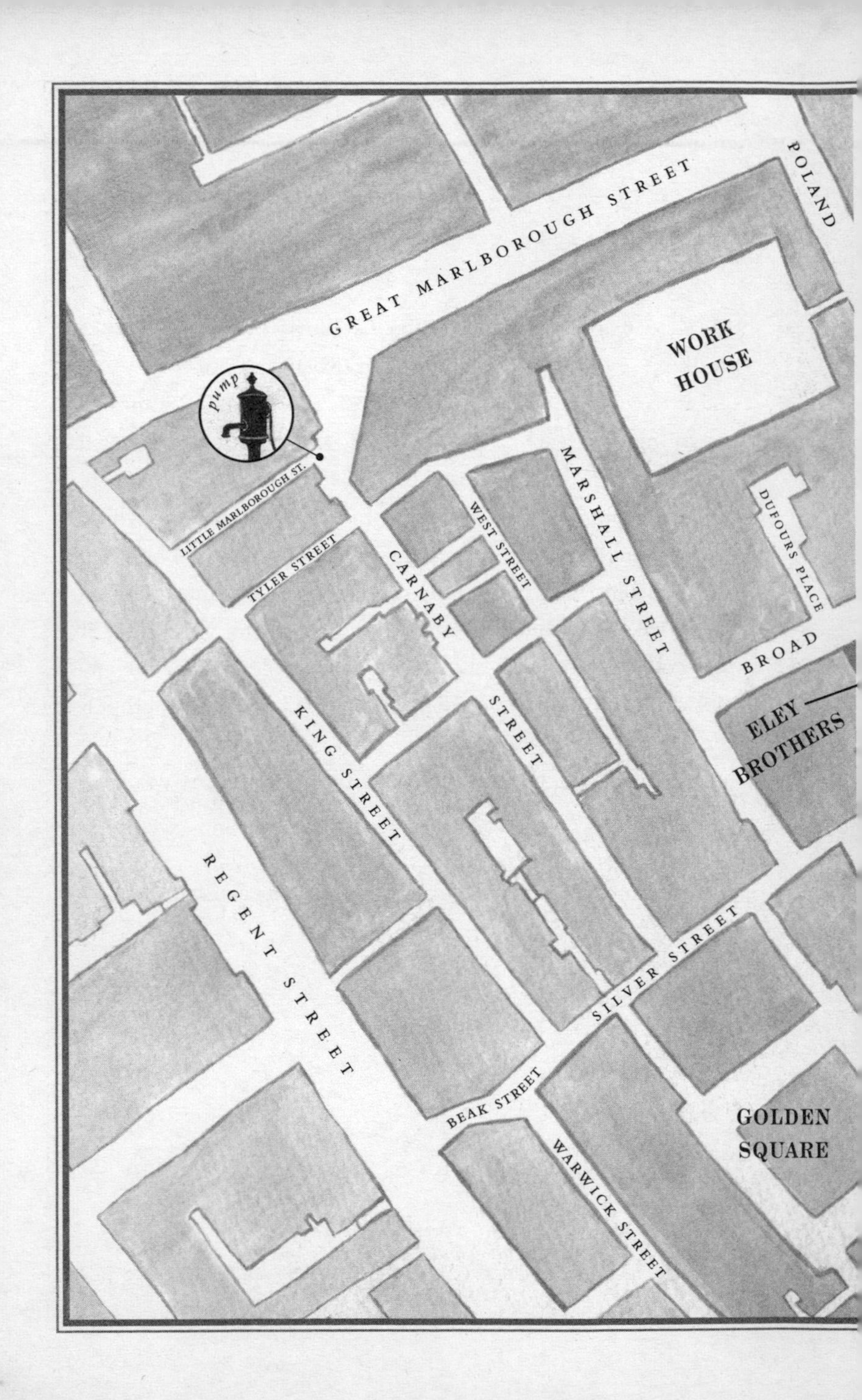
GREAT MARLBOROUGH STREET
POLAND
WORK HOUSE
pump
LITTLE MARLBOROUGH ST.
TYLER STREET
CARNABY
WEST STREET
MARSHALL STREET
DUFOURS PLACE
BROAD
ELEY BROTHERS
KING STREET
STREET
REGENT STREET
SILVER STREET
BEAK STREET
WARWICK STREET
GOLDEN SQUARE

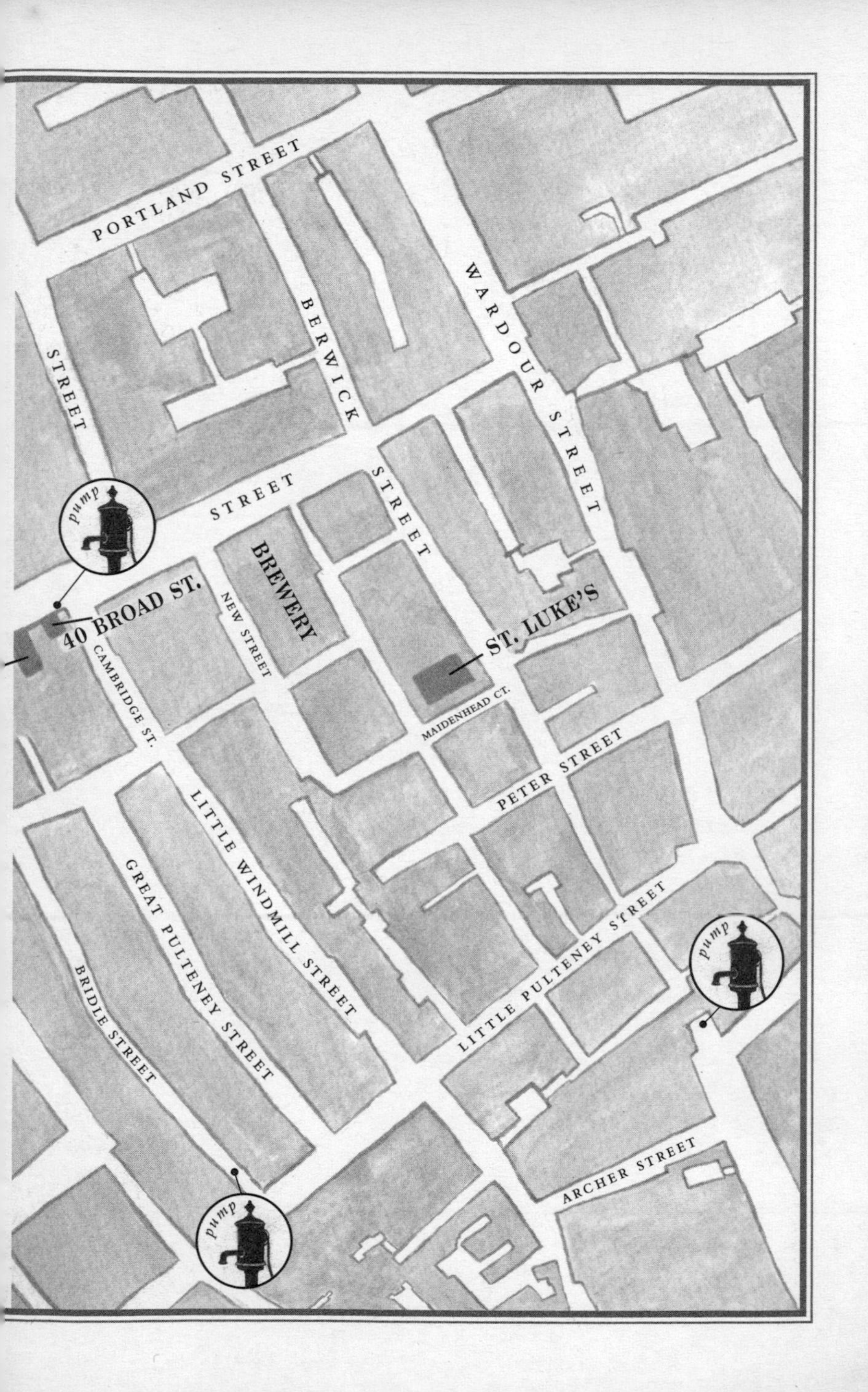
PORTLAND STREET
BERWICK
WARDOUR STREET
STREET
STREET
pump
STREET
BREWERY
40 BROAD ST.
NEW STREET
ST. LUKE'S
CAMBRIDGE ST.
MAIDENHEAD CT.
PETER STREET
LITTLE WINDMILL STREET
GREAT PULTENEY STREET
LITTLE PULTENEY STREET
pump
BRIDLE STREET
ARCHER STREET
pump

This is a story with four protagonists: a deadly bacterium, a vast city, and two gifted but very different men. One dark week a hundred fifty years ago, in the midst of great terror and human suffering, their lives collided on London's Broad Street, on the western edge of Soho.

This book is an attempt to tell the story of that collision in a way that does justice to the multiple scales of existence that helped bring it about: from the invisible kingdom of microscopic bacteria, to the tragedy and courage and camaraderie of individual lives, to the cultural realm of ideas and ideologies, all the way up to the sprawling metropolis of London itself. It is the story of a map that lies at the intersection of all those different vectors, a map created to help make sense of an experience that defied human understanding. It is also a case study in how change happens in human society, the turbulent way in which wrong or ineffectual ideas are overthrown by better ones. More than anything else, though, it is an argument for seeing that terrible week as one of the defining moments in the invention of modern life.

THE GHOST MAP

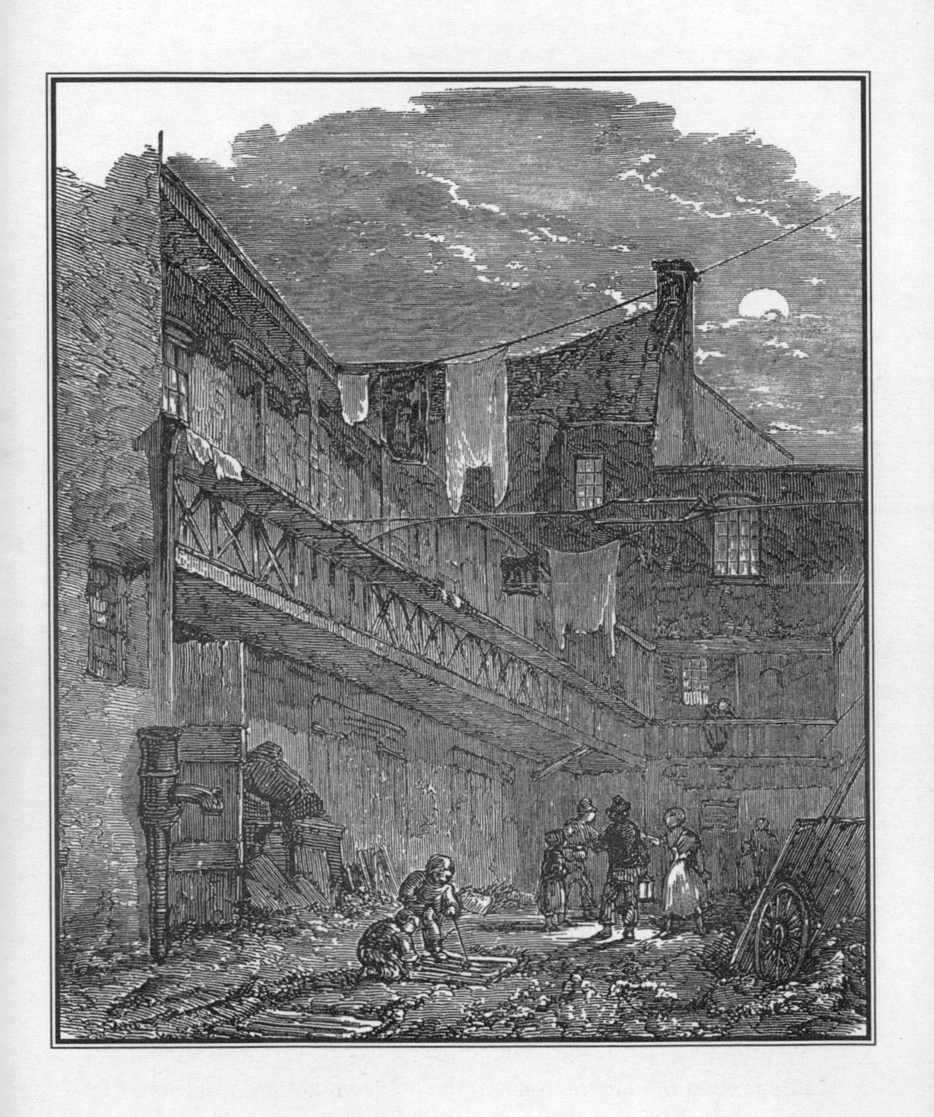

Monday, August 28

THE NIGHT-SOIL MEN

IT IS AUGUST 1854, AND LONDON IS A CITY OF SCAVENGERS. Just the names alone read now like some kind of exotic zoological catalogue: bone-pickers, rag-gatherers, pure-finders, dredgermen, mud-larks, sewer-hunters, dustmen, night-soil men, bunters, toshers, shoremen. These were the London underclasses, at least a hundred thousand strong. So immense were their numbers that had the scavengers broken off and formed their own city, it would have been the fifth-largest in all of England. But the diversity and precision of their routines were more remarkable than their sheer number. Early risers strolling along the Thames would see the toshers wading through the muck of low tide, dressed almost comically in flowing velveteen coats, their oversized pockets filled with stray bits of copper recovered from the water's edge. The toshers walked with a lantern strapped to their chest to help them see in the predawn gloom, and carried an eight-foot-long pole that they used to test the ground in front of them, and to pull themselves out when they stumbled into a quagmire. The pole and the eerie glow of the lantern through the robes gave them the look of ragged wizards, scouring the foul river's edge for magic coins. Beside them fluttered the mud-larks, often children, dressed in tatters and content to scavenge all the waste that the toshers rejected as below their standards: lumps of coal, old wood, scraps of rope.

Above the river, in the streets of the city, the pure-finders eked out a living by collecting dog shit (colloquially called "pure") while the bone-pickers foraged for carcasses of any stripe. Below ground, in the cramped but growing network of tunnels beneath London's streets, the sewer-hunters slogged through the flowing waste of the metropolis. Every few months, an unusually dense pocket of methane gas would be ignited by one of their kerosene lamps and the hapless soul would be incinerated twenty feet below ground, in a river of raw sewage.

The scavengers, in other words, lived in a world of excrement and death. Dickens began his last great novel, *Our Mutual Friend,* with a father-daughter team of toshers stumbling across a corpse floating in the Thames, whose coins they solemnly pocket. "What world does a dead man belong to?" the father asks rhetorically, when chided by a fellow tosher for stealing from a corpse. "'Tother world. What world does money belong to? This world." Dickens' unspoken point is that the two worlds, the dead and the living, have begun to coexist in these marginal spaces. The bustling commerce of the great city has conjured up its opposite, a ghost class that somehow mimics the status markers and value calculations of the material world. Consider the haunting precision of the bone-pickers' daily routine, as captured in Henry Mayhew's pioneering 1844 work, *London Labour and the London Poor*:

> It usually takes the bone-picker from seven to nine hours to go over his rounds, during which time he travels from 20 to 30 miles with a quarter to a half hundredweight on his back. In the summer he usually reaches home about eleven of the day, and in the winter about one or two. On his return home he proceeds to sort the contents of his bag. He separates the rags from the bones, and these again from the old metal (if he be luckly enough to have found any). He divides the rags into various lots, according as they are white or coloured; and if he have picked up any pieces of canvas or sacking, he makes these also into a separate parcel. When he has finished the sorting he takes his several lots to the ragshop or the marine-store dealer, and realizes upon them whatever they may be worth. For the white rags he gets from 2d. to 3d. per pound, according as they are clean or soiled. The white rags are very difficult to be found; they are mostly very dirty, and are therefore sold with the coloured ones at the rate of about 5 lbs. for 2d.

The homeless continue to haunt today's postindustrial cities, but they rarely display the professional clarity of the bone-picker's impromptu trade, for two primary reasons. First, minimum wages and government assistance are now substantial enough that it no longer makes economic sense to eke out a living as a scavenger. (Where wages remain depressed, scavenging remains a vital occupation; witness the *perpendadores* of Mexico City.) The bone collector's trade has also declined because most modern cities possess elaborate systems for managing the waste generated by their inhabitants. (In fact, the closest American equivalent to the Victorian scavengers—the aluminum-can collectors you sometimes see hovering outside supermarkets—rely on precisely those waste-management systems for their paycheck.) But London in 1854 was a Victorian metropolis trying to make do with an Elizabethan public infrastructure. The city was vast even by today's standards, with two and a half million people crammed inside a thirty-mile circumference. But most of the techniques for managing that kind of population density that we now take for granted—recycling centers, public-health departments, safe sewage removal—hadn't been invented yet.

And so the city itself improvised a response—an unplanned, organic response, to be sure, but at the same time a response that was precisely contoured to the community's waste-removal needs. As the garbage and excrement grew, an underground market for refuse developed, with hooks into established trades. Specialists emerged, each dutifully carting goods to the appropriate site in the official market: the bone collectors selling their goods to the bone-boilers, the pure-finders selling their dog shit to tanners, who used the "pure" to rid their leather goods of the lime they had soaked in for weeks to remove animal hair. (A process widely considered to be, as one tanner put it, "the most disagreeable in the whole range of manufacture.")

We're naturally inclined to consider these scavengers tragic figures, and to fulminate against a system that allowed so many thousands to eke out a living by foraging through human waste. In many ways, this is the correct response. (It was, to be sure, the response of the great crusaders of the age, among them Dickens and Mayhew.) But such social outrage should be accompanied by a measure of wonder and respect: without any central planner coordinating their actions, without any education at all, this itinerant underclass managed to conjure up an entire system for processing and sorting the waste generated by two million people. The great contribution usually ascribed to Mayhew's *London Labour* is simply his willingness to see and record the details of these impoverished lives. But just as valuable was the insight that came out of that bookkeeping, once he had run the numbers: far from being unproductive vagabonds, Mayhew discovered, these people were actually performing an essential function for their community. "The removal of the refuse of a large town," he wrote, "is, perhaps, one of the most important of social operations." And the scavengers of Victorian London weren't just getting rid of that refuse—they were recycling it.

WASTE RECYCLING IS USUALLY ASSUMED TO BE AN INVENTION of the environmental movement, as modern as the blue plastic bags we now fill with detergent bottles and soda cans. But it is an ancient art. Composting pits were used by the citizens of Knossos in Crete four thousand years ago. Much of medieval Rome was built out of materials pilfered from the crumbling ruins of the imperial city. (Before it was a tourist landmark, the Colosseum served as a de facto quarry.) Waste recycling—in the form of composting and manure spreading—played a crucial role in the explosive growth of medieval European towns. High-density collections of human beings, by definition, require significant energy inputs to be sustainable, starting with reliable supplies of food. The towns of the Middle Ages lacked highways and container ships to bring them sustenance, and so their population sizes were limited by the fecundity of the land around them. If the land could grow only enough food to sustain five thousand people, then five thousand people became the ceiling. But by plowing their organic waste back into the earth, the early medieval towns increased the productivity of the soil, thus raising the population ceiling, thereby creating more waste—and increasingly fertile soil. This feedback loop transformed the boggy expanses of the Low Countries, which had historically been incapable of sustaining anything more complex than isolated bands of fishermen, into some of the most productive soils in all of Europe. To this day, the Netherlands has the highest population density of any country in the world.

Waste recycling turns out to be a hallmark of almost all complex systems, whether the man-made ecosystems of urban life, or the microscopic economies of the cell. Our bones are themselves the result of a recycling scheme pioneered by natural selection billions of years ago. All nucleated organisms generate excess calcium as a waste product. Since at least the Cambrian times, organisms have accumulated those calcium reserves, and put them to good use: building shells, teeth, skeletons. Your ability to walk upright is due to evolution's knack for recycling its toxic waste.

Waste recycling is a crucial attribute of the earth's most diverse ecosystems. We value tropical rain forests because they squander so little of the energy supplied by the sun, thanks to their vast, interlocked system of organisms exploiting every tiny niche of the nutrient cycle. The cherished diversity of the rain-forest ecosystem is not just a quaint case of biological multiculturalism. The diversity of the system is precisely why rain forests do such a brilliant job of capturing the energy that flows through them: one organism captures a certain amount of energy, but in processing that energy, it generates waste. In an efficient system, that waste becomes a new source of energy for another creature in the chain. (That efficiency is one of the reasons why clearing the rain forests is such a shortsighted move: the nutrient cycles in their ecosystems are so tight that the soil is usually very poor for farming: all the available energy has been captured on its way down to the forest floor.)

Coral reefs display a comparable knack for waste management. Corals live in a symbiotic alliance with tiny algae called zooxanthellae. Thanks to photosynthesis, the algae capture sunlight and use it to turn carbon dioxide into organic carbon, with oxygen as a waste product of the process. The coral then uses the oxygen in its own metabolic cycle. Because we're aerobic creatures ourselves, we tend not to think of oxygen as a waste product, but from the point of view of the algae, that's precisely what it is: a useless substance discharged as part of its metabolic cycle. The coral itself produces waste in the form of carbon dioxide, nitrates, and phosphates, all of which help the algae to grow. That tight waste-recycling chain is one of the primary reasons coral reefs are able to support such a dense and diverse population of creatures, despite residing in tropical waters, which are generally nutrient-poor. They are the cities of the sea.

There can be many causes behind extreme population density—whether the population is made up of angelfish or spider monkeys or humans—but without efficient forms of waste recycling, those dense concentrations of life can't survive for long. Most of that recycling work, in both remote tropical rain forests and urban centers, takes place at the microbial level. Without the bacteria-driven processes of decomposition, the earth would have been overrun by offal and carcasses eons ago, and the life-sustaining envelope of the earth's atmosphere would be closer to the uninhabitable, acidic surface of Venus. If some rogue virus wiped out every single mammal on the planet, life on earth would proceed, largely unaffected by the loss. But if the bacteria disappeared overnight, all life on the planet would be extinguished within a matter of years.

You couldn't see those microbial scavengers at work in Victorian London, and the great majority of scientists—not to mention laypeople—had no idea that the world was in fact teeming with tiny organisms that made their lives possible. But you could detect them through another sensory channel: smell. No extended description of London from that period failed to mention the stench of the city. Some of that stench came from the burning of industrial fuels, but the most objectionable smells—the ones that ultimately helped prod an entire public-health infrastructure into place—came from the steady, relentless work of bacteria decomposing organic matter. Those deadly pockets of methane in the sewers were themselves produced by the millions of microorganisms diligently recycling human dung into a microbial biomass, with a variety of gases released as waste products. You can think of those fiery, underground explosions as a kind of skirmish between two different kinds of scavenger: sewer-hunter versus bacterium—living on different scales but nonetheless battling for the same territory.

But in that late summer of 1854, as the toshers and the mud-larks and the bone collectors made their rounds, London was headed toward another, even more terrifying, battle between microbe and man. By the time it was over, it would prove as deadly as any in the city's history.

LONDON'S UNDERGROUND MARKET OF SCAVENGING HAD ITS own system of rank and privilege, and near the top were the night-soil men. Like the beloved chimney sweeps of *Mary Poppins,* the night-soil men worked as independent contractors at the very edge of the legitimate economy, though their labor was significantly more revolting than the foraging of the mud-larks and toshers. City landlords hired the men to remove the "night soil" from the overflowing cesspools of their buildings. The collecting of human excrement was a venerable occupation; in medieval times they were called "rakers" and "gong-fermors," and they played an indispensable role in the waste-recycling system that helped London grow into a true metropolis, by selling the waste to farmers outside the city walls. (Later entrepreneurs hit upon a technique for extracting nitrogen from the ordure that could be reused in the manufacture of gunpowder.) While the rakers and their descendants made a good wage, the work conditions could be deadly: in 1326, an ill-fated laborer by the name of Richard the Raker fell into a cesspool and literally drowned in human shit.

By the nineteenth century, the night-soil men had evolved a precise choreography for their labors. They worked the graveyard shift, between midnight and five a.m., in teams of four: a "ropeman," a "holeman," and two "tubmen." The team would affix lanterns at the edge of the cesspit, then remove the floorboards or stone covering it, sometimes with a pickax. If the waste had accumulated high enough, the ropeman and holeman would begin by scooping it out with the tub. Eventually, as more night soil was removed, the men would lower a ladder down and the holeman would descend into the pit and scoop waste into his tub. The ropeman would help pull up each full tub, and pass it along to the tubmen who emptied the waste into their carts. It was standard practice for the night-soil men to be offered a bottle of gin for their labors. As one reported to Mayhew: "I should say that there's been a bottle of gin drunk at the clearing of every two, ay, and more than every two, out of three cesspools emptied in London; and now that I come to think on it, I should say that's been the case with three out of every four."

The work was foul, but the pay was good. Too good, as it turned out. Thanks to its geographic protection from invasion, London had become the most sprawling of European cities, expanding far beyond its Roman walls. (The other great metropolis of the nineteenth century, Paris, had almost the same population squeezed into half the geographic area.) For the night-soil men, that sprawl meant longer transport times—open farmland was now often ten miles away—which drove the price of their removing waste upward. By the Victorian era, the night-soil men were charging a shilling a cesspool, wages that were at least twice that of the average skilled laborer. For many Londoners, the financial cost of removing waste exceeded the environmental cost of just letting it accumulate—particularly for landlords, who often didn't live on top of these overflowing cesspools. Sights like this one, reported by a civil engineer hired to survey two houses under repair in the 1840s, became commonplace: "I found whole areas of the cellars of both houses were full of nightsoil to the depth of three feet, which had been permitted for years to accumulate from the overflow of the cesspools. . . . Upon passing through the passage of the first house I found the yard covered in nightsoil, from the overflowing of the privy to the depth of nearly six inches and bricks were placed to enable the inmates to get across dryshod." Another account describes a dustheap in Spitalfields, in the heart of the East End: "a heap of dung the size of a tolerably large house, and an artificial pond into which the content of cesspits are thrown. The contents are allowed to desiccate in the open air, and they are frequently stirred for that purpose." Mayhew described this grotesque scene in an article published in the London *Morning Chronicle* in 1849 that surveyed the ground zero of that year's cholera outbreak:

> We then journeyed on to London-street. . . . In No. 1 of this street the cholera first appeared seventeen years ago, and spread up it with fearful virulence; but this year it appeared at the opposite end, and ran down it with like severity. As we passed along the reeking banks of the sewer, the sun shone upon a narrow slip of the water. In the

> bright light it appeared the colour of strong green tea, and positively looked as solid as black marble in the shadow—indeed, it was more like watery mud than muddy water; and yet we were assured this was the only water the wretched inhabitants had to drink. As we gazed in horror at it, we saw drains and sewers emptying their filthy contents into it; we saw a whole tier of doorless privies in the open road, common to men and women, built over it; we heard bucket after bucket of filth splash into it; and the limbs of the vagrant boys bathing in it seemed by pure force of contrast, white as Parian marble. And yet, as we stood doubting the fearful statement, we saw a little child, from one of the galleries opposite, lower a tin can with a rope to fill a large bucket that stood beside her. In each of the balconies that hung over the stream the self-same tub was to be seen in which the inhabitants put the mucky liquid to stand, so that they may, after it has rested for a day or two, skim the fluid from the solid particles of filth, pollution, and disease. As the little thing dangled her tin cup as gently as possible into the stream, a bucket of night-soil was poured down from the next gallery.

Victorian London had its postcard wonders, to be sure—the Crystal Palace, Trafalgar Square, the new additions to Westminster Palace. But it also had wonders of a different order, no less remarkable: artificial ponds of raw sewage, dung heaps the size of houses.

The elevated wage of the night-soil men wasn't the only culprit behind this rising tide of excrement. The runaway popularity of the water closet heightened the crisis. A water-flushing device had been invented in the late sixteenth century by Sir John Harington, who actually installed a functioning version for his godmother, Queen Elizabeth, at Richmond Palace. But the device didn't take off until the late 1700s, when a watchmaker named Alexander Cummings and a cabinetmaker named Joseph Bramah filed for two separate patents on an improved version of Harington's design. Bramah went on to build a profitable business installing water closets in the homes of the well-to-do. According to one survey, water-closet installations had increased tenfold in the period between 1824 and 1844. Another spike happened after a manufacturer named George Jennings installed water closets for public use in Hyde Park during the Great Exhibition of 1851. An estimated 827,000 visitors used them. The visitors no doubt marveled at the Exhibition's spectacular display of global culture and modern engineering, but for many the most astonishing experience was just sitting on a working toilet for the first time.

Water closets were a tremendous breakthrough as far as quality of life was concerned, but they had a disastrous effect on the city's sewage problem. Without a functioning sewer system to connect to, most WCs simply flushed their contents into existing cesspools, greatly increasing their tendency to overflow. According to one estimate, the average London household used 160 gallons of water a day in 1850. By 1856, thanks to the runaway success of the water closet, they were using 244 gallons.

But the single most important factor driving London's waste-removal crisis was a matter of simple demography: the number of people generating waste had almost tripled in the space of fifty years. In the 1851 census, London had a population of 2.4 million people, making it the most populous city on the planet, up from around a million at the turn of the century. Even with a modern civic infrastructure, that kind of explosive growth is difficult to manage. But without infrastructure, two million people suddenly forced to share ninety square miles of space wasn't just a disaster waiting to happen—it was a kind of permanent, rolling disaster, a vast organism destroying itself by laying waste to its habitat. Five hundred years after the fact, London was slowly re-creating the horrific demise of Richard the Raker: it was drowning in its own filth.

ALL OF THOSE HUMAN LIVES CROWDED TOGETHER HAD AN inevitable repercussion: a surge in corpses. In the early 1840s, a twenty-three-year-old Prussian named Friedrich Engels embarked on a scouting mission for his industrialist father that inspired both a classic text of urban sociology and the modern Socialist movement. Of his experiences in London, Engels wrote:

> The corpses [of the poor] have no better fate than the carcasses of animals. The pauper burial ground at St Bride's is a piece of open marshland which has been used since Charles II's day and there are heaps of bones all over the place. Every Wednesday the remains of dead paupers are thrown in to a hole which is 14 feet deep. A clergyman gabbles through the burial service and then the grave is filled with loose soil. On the following Wednesday the ground is opened again and this goes on until it is completely full. The whole neighborhood is infected from the dreadful stench.

One privately run burial ground in Islington had packed 80,000 corpses into an area designed to hold roughly three thousand. A gravedigger there reported to the *Times* of London that he had been "up to my knees in human flesh, jumping on the bodies, so as to cram them in the least possible space at the bottom of the graves, in which fresh bodies were afterwards placed."

Dickens buries the mysterious opium-addicted law-writer who

overdoses near the beginning of *Bleak House* in a comparably grim setting, inspiring one of the book's most famous, and famously impassioned, outbursts:

> a hemmed-in churchyard, pestiferous and obscene, whence malignant diseases are communicated to the bodies of our dear brothers and sisters who have not departed. . . . With houses looking on, on every side, save where a reeking little tunnel of a court gives access to the iron gate—with every villainy of life in action close on death, and every poisonous element of death in action close on life—here, they lower our dear brother down a foot or two: here, sow him in corruption, to be raised in corruption: an avenging ghost at many a sick-bedside: a shameful testimony to future ages, how civilization and barbarism walked this boastful island together.

To read those last sentences is to experience the birth of what would become a dominant rhetorical mode of twentieth-century thought, a way of making sense of the high-tech carnage of the Great War, or the Taylorite efficiencies of the concentration camps. The social theorist Walter Benjamin reworked Dickens' original slogan in his enigmatic masterpiece "Theses on the Philosophy of History," written as the scourge of fascism was enveloping Europe: "There is no document of civilization that is not also a document of barbarism."

The opposition between civilization and barbarism was practically as old as the walled city itself. (As soon as there were gates, there were barbarians ready to storm them.) But Engels and Dickens suggested a new twist: that the advance of civilization produced barbarity as an unavoidable waste product, as essential to its metabolism as the gleaming spires and cultivated thought of polite society. The barbarians weren't storming the gates. They were being bred from within. Marx took that insight, wrapped it in Hegel's dialectics, and transformed the twentieth century. But the idea itself sprang out of a certain kind of lived experience—on the ground, as the activists still like to say. It came, in part, from seeing human beings buried in conditions that defiled both the dead and the living.

But in one crucial sense Dickens and Engels had it wrong. However gruesome the sight of the burial ground was, the corpses themselves were not likely spreading "malignant diseases." The stench was offensive enough, but it was not "infecting" anyone. A mass grave of decomposing bodies was an affront to both the senses and to personal dignity, but the smell it emitted was not a public-health risk. No one died of stench in Victorian London. But tens of thousands died because the fear of stench blinded them to the true perils of the city, and drove them to implement a series of wrongheaded reforms that only made the crisis worse. Dickens and Engels were not alone; practically the entire medical and political establishment fell into the same deadly error: everyone from Florence Nightingale to the pioneering reformer Edwin Chadwick to the editors of *The Lancet* to Queen Victoria herself. The history of knowledge conventionally focuses on breakthrough ideas and conceptual leaps. But the blind spots on the map, the dark continents of error and prejudice, carry their own mystery as well. How could so many intelligent people be so grievously wrong for such an extended period of time? How could they ignore so much overwhelming evidence that contradicted their most basic theories? These questions, too, deserve their own discipline—the sociology of error.

The fear of death's contamination can sometimes last for centuries. In the middle of the Great Plague of 1665, the Earl of Craven purchased a block of land in a semirural area to the west of central London called Soho Field. He built thirty-six small houses "for the reception of poor and miserable objects" suffering from plague. The rest of the land was used as a mass grave. Each night, the death carts would empty dozens of corpses into the earth. By some estimates, over four thousand plague-infected bodies were buried there in a matter of months. Nearby residents gave it the appropriately macabre-sounding name of "Earl of Craven's pest-field," or "Craven's field" for short. For two generations, no one dared erect a foundation in the land for fear of infection. Eventually, the city's inexorable drive for shelter won out over its fear of disease, and the pesthouse fields became the fashionable district of Golden Square, populated largely by aristocrats and Huguenot immigrants. For another century, the skeletons lay undisturbed beneath the churn of city commerce, until late summer of 1854, when another outbreak came to Golden Square and brought those grim souls back to haunt their final resting grounds once more.

CRAVEN'S FIELD ASIDE, SOHO IN THE DECADES AFTER THE plague quickly became one of London's most fashionable neighborhoods. Almost a hundred titled families lived there in the 1690s. In 1717, the Prince and Princess of Wales set up residence in Leicester House in Soho. Golden Square itself had been built out with elegant Georgian townhouses, a haven from the tumult of Piccadilly Circus several blocks to the south. But by the middle of the eighteenth century, the elites continued their inexorable march westward, building even grander estates and townhouses in the burgeoning new neighborhood of Mayfair. By 1740, there were only twenty titled residents left. A new kind of Soho native began to appear, best embodied by the son of a hosier who was born at 28 Broad in 1757, a talented and troubled child by the name of William Blake, who would go on to be one of England's greatest poets and artists. In his late twenties, he returned to Soho and opened a printing shop next door to his late father's shop, now run by his brother. Another Blake brother opened a bakery across the road at 29 Broad shortly thereafter, and so for a few years, the Blake family had a mini-empire growing on Broad Street, with three separate businesses on the same block.

The mix of artistic vision and entrepreneurial spirit would define the area for several generations. As the city grew increasingly industrial, and as the old money emptied out, the neighborhood became grittier; landlords invariably broke up the old townhouses into separate flats; courtyards between buildings filled up with impromptu junkyards, stables, jury-rigged extensions. Dickens described it best in *Nicholas Nickleby*:

> In that quarter of London in which Golden Square is situated, there is a bygone, faded, tumble-down street, with two irregular rows of tall meagre houses, which seem to have stared each other out of countenance years ago. The very chimneys appear to have grown dismal and melancholy from having had nothing better to look at than the chimneys over the way. . . . To judge from the size of the houses, they have been, at one time, tenanted by persons of better condition than their present occupants; but they are now let off, by the week, in floors or rooms, and every door has almost as many plates or bell-handles as there are apartments within. The windows are, for the same reason, sufficiently diversified in appearance, being ornamented with every variety of common blind and curtain that can easily be imagined; which every doorway is blocked up, and rendered nearly impassable, by a motley collection of children and porter pots of all sizes, from the baby in arms and the half-pint pot, to the full-grown girl and half-gallon can.

By 1851, the subdistrict of Berwick Street on the west side of Soho was the most densely populated of all 135 subdistricts that made up Greater London, with 432 people to the acre. (Even with its skyscrapers, Manhattan today only houses around 100 per acre.) The parish of St. Luke's in Soho had thirty houses per acre. In Kensington, by contrast, the number per acre was two.

But despite—or perhaps because of—the increasingly crowded and unsanitary conditions, the neighborhood was a hotbed of creativity. The list of poets and musicians and sculptors and philosophers who lived in Soho during this period reads like an index to a textbook on Enlightenment-era British culture. Edmund Burke, Fanny Burney, Percy Shelley, William Hogarth—all were Soho residents at various points in their lives. Leopold Mozart leased a flat on Frith Street while visiting with his son, the eight-year-old prodigy Wolfgang, in 1764. Franz Liszt and Richard Wagner also stayed in the neighborhood when visiting London in 1839–1840.

"New ideas need old buildings," Jane Jacobs once wrote, and the maxim applies perfectly to Soho around the dawn of the Industrial Age: a class of visionaries and eccentrics and radicals living in the disintegrating shells that had been abandoned a century ago by the well-to-do. The trope is familiar to us by now—artists and renegades appropriate a decaying neighborhood, even relish the decay—but it was a new pattern of urban settlement when Blake and Hogarth and Shelley first made their homes along the crowded streets of Soho. They seem to have been energized by the squalor, not appalled by it. Here is a description of one typical residence on Dean Street, penned in the early 1850s:

> [The flat] has two rooms, the one with the view of the street being the drawing-room, behind it the bedroom. There is not one piece of

> good, solid furniture in the entire flat. Everything is broken, tattered and torn, finger-thick dust everywhere, and everything in the greatest disorder. . . . When you enter the . . . flat, your sight is dimmed by tobacco and coal smoke so that you grope around at first as if you were in a cave, until your eyes get used to the fumes and, as in a fog, you gradually notice a few objects. Everything is dirty, everything covered with dust; it is dangerous to sit down.

Living in this two-room attic were seven individuals: a Prussian immigrant couple, their four children, and a maid. (Apparently a maid with an aversion to dusting.) Yet somehow these cramped, tattered quarters did not noticeably hinder the husband's productivity, though one can easily see why he developed such a fondness for the Reading Room at the British Museum. The husband, you see, was a thirty-something radical by the name of Karl Marx.

By the time Marx got to Soho, the neighborhood had turned itself into the kind of classic mixed-use, economically diverse neighborhood that today's "new urbanists" celebrate as the bedrock of successful cities: two-to-four-story residential buildings with storefronts at nearly every address, interlaced with the occasional larger commercial space. (Unlike the typical new urbanist environment, however, Soho also had its share of industry: slaughterhouses, manufacturing plants, tripe boilers.) The neighborhood's residents were poor, almost destitute, by the standards of today's industrialized nations, though by Victorian standards they were a mix of the working poor and the entrepreneurial middle class. (By mud-lark standards, of course, they were loaded.) But Soho was something of an anomaly in the otherwise prosperous West End of the city: an island of working poverty and foul-smelling industry surrounded by the opulent townhouses of Mayfair and Kensington.

This economic discontinuity is still encoded in the physical layout of the streets around Soho. The western border of the neighborhood is defined by the wide avenue of Regent Street, with its gleaming white commercial façades. West of Regent Street you enter the tony enclave of Mayfair, posh to this day. But somehow the nonstop traffic and bustle of Regent Street is almost imperceptible from the smaller lanes and alleys of western Soho, largely because there are very few conduits that open directly onto Regent Street. Walking around the neighborhood, it feels almost as if a barricade has been erected, keeping you from reaching the prominent avenue that you know is only a few feet away. And indeed, the street layout was explicitly designed to serve as a barricade. When John Nash designed Regent Street to connect Marylebone Park with the Prince Regent's new home at Carlton House, he planned the thoroughfare as a kind of *cordon sanitaire* separating the well-to-do of Mayfair from the growing working-class community of Soho. Nash's explicit intention was to create "a complete separation between the streets occupied by the Nobility and Gentry, and the narrower Streets and meaner houses occupied by mechanics and the trading part of the community. . . . My purpose was that the new street should cross the eastern entrance to all the streets occupied by the higher classes and to leave out to the east all the bad streets."

This social topography would play a pivotal role in the events that unfolded in the late summer of 1854, when a terrible scourge struck Soho but left the surrounding neighborhoods utterly unharmed. That selective attack appeared to confirm every elitist cliché in the book: the plague attacking the debauched and the destitute, while passing over the better sort that lived only blocks away. Of course the plague had devastated the "meaner houses" and "bad streets"; anyone who had visited those squalid blocks would have seen it coming.

Poverty and depravity and low breeding created an environment where disease prospered, as anyone of good social standing would tell you. That's why they'd built barricades in the first place.

But on the wrong side of Regent Street, behind the barricade, the tradesmen and the mechanics managed to get by in the mean houses of Soho. The neighborhood was a veritable engine of local commerce, with almost every residence housing some kind of small business. The assortment of storefronts generally sounds quaint to the modern ear. There were the grocers and bakeries that wouldn't be out of place in an urban center today; but there were also the machinists and mineral teeth manufacturers doing business beside them. In August of 1854, walking down Broad Street, a block north of Golden Square, one would have encountered, in progression: a grocer, a bonnet maker, a baker, a grocer, a saddle-tree manufacturer, an engraver, and ironmonger, a trimming seller, a percussion-cap manufacturer, a wardrobe dealer, a boot-tree manufacturer, and a pub, The Newcastle-on-Tyne. In terms of professions, tailors outnumbered any other trade by a relatively wide margin. After the tailors, at roughly the same number, were the shoemakers, domestic servants, masons, shopkeepers, and dressmakers.

Sometime in the late 1840s, a London policeman named Thomas Lewis and his wife moved into 40 Broad Street, one door up from the pub. It was an eleven-room house that had originally been designed to hold a single family and a handful of servants. Now it contained twenty inhabitants. These were spacious accommodations for a part of the city where most houses averaged five occupants per room. Thomas and Sarah Lewis lived in the parlor at 40 Broad, first with their little boy, a sickly child who died at ten months. In March of 1854, Sarah Lewis gave birth to a girl, who possessed, from the beginning, a more promising constitution than her late brother. Sarah

Lewis had been unable to breast-feed the infant on account of health problems of her own, but she had fed her daughter ground rice and milk from a bottle. The little girl had suffered a few bouts of illness in her second month, but was relatively healthy for most of the summer.

A few mysteries remain about this second Lewis infant, details scattered by the chance winds of history. We do not know her name, for instance. We do not know what series of events led to her contracting cholera in late August of 1854, at not even six months old. For almost twenty months, the disease had been flaring up in certain quarters of London, having last appeared during the revolutionary years of 1848–1849. (Plagues and political unrest have a long history of following the same cycles.) But most of the cholera outbreaks in 1854 were located south of the Thames. The Golden Square area had been largely spared.

On the twenty-eighth of August, all that changed. At around six a.m., while the rest of the city struggled for a few final minutes of sleep at the end of an oppressively hot summer night, the Lewis infant began vomiting and emitting watery, green stools that carried a pungent smell. Sarah Lewis sent for a local doctor, William Rogers, who maintained a practice a few blocks away, on Berners Street. As she waited for the doctor's arrival, Sarah soaked the soiled cloth diapers in a bucket of tepid water. In the rare moments when her little girl caught a few minutes of sleep, Sarah Lewis crept down to the cellar at 40 Broad and tossed the fouled water in the cesspool that lay at the front of the house.

That is how it began.

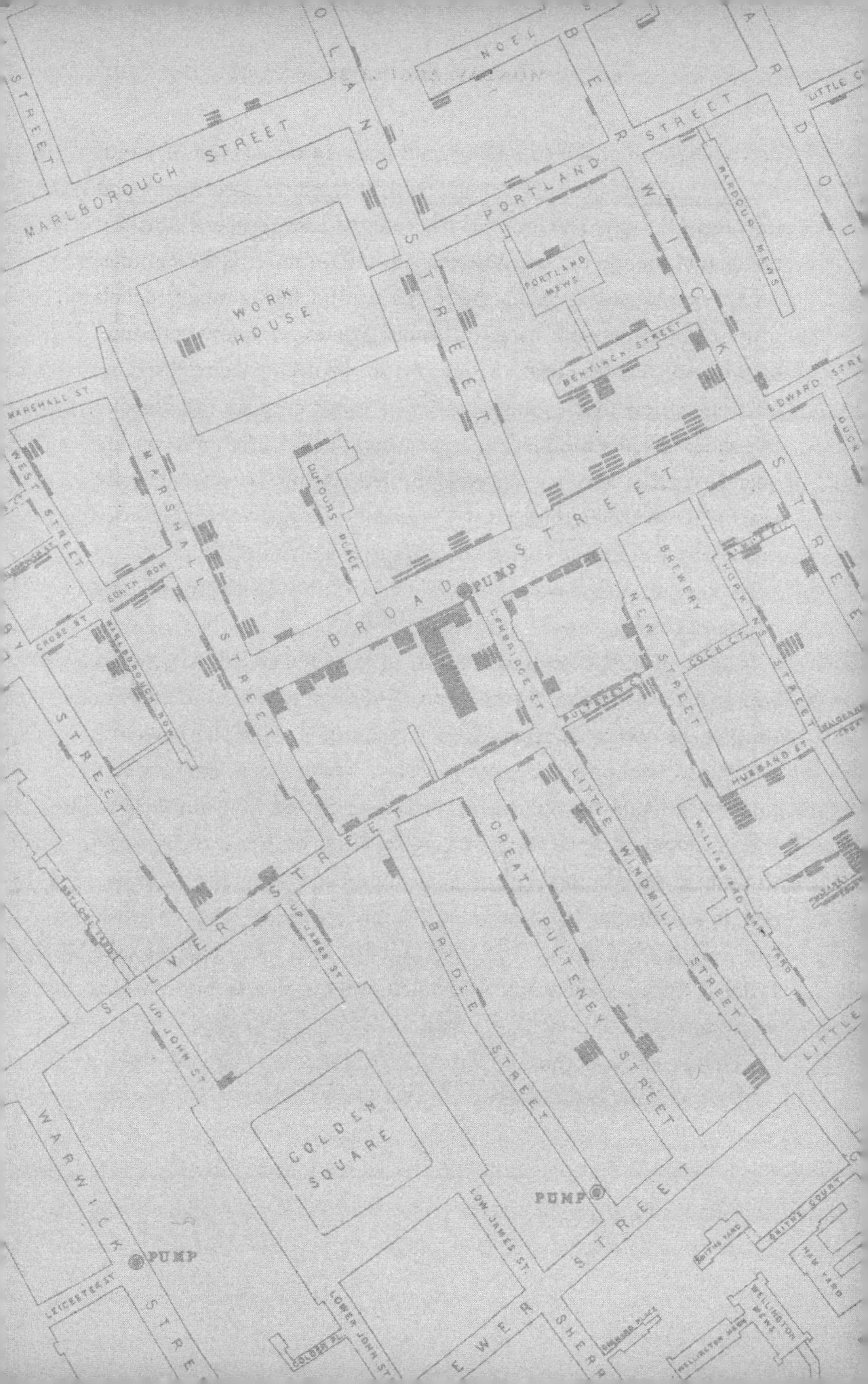

MARLBOROUGH STREET
STREET
NOEL
PORTLAND STREET
PORTLAND MEWS
WORK HOUSE
MARSHALL ST
DUFOURS PLACE
EDWARD STREET
BROAD STREET
PUMP
CAMBRIDGE ST
BREWERY
NEW STREET
WEST STREET
CROSS ST
MARSHALL STREET
STREET
LITTLE WINDMILL STREET
GREAT PULTENEY STREET
BRIDLE STREET
SILVER STREET
UP JAMES ST
UP JOHN ST
GOLDEN SQUARE
PUMP
LOW JAMES ST
STREET
WARWICK STREET
PUMP
LEICESTER ST
LOWER JOHN ST
WELLINGTON MEWS
SMITHS COURT
NEW YARD

HENRY WHITEHEAD

Saturday, September 2

EYES SUNK, LIPS DARK BLUE

For two days after the Lewis baby fell ill, life in Golden Square carried on with its normal clamor. In nearby Soho Square, an affable clergyman named Henry Whitehead took leave of the boarding room he shared with his brother and embarked on his morning stroll to St. Luke's Church on Berwick Street, where he had been appointed assistant curate. Only twenty-eight years old, Whitehead had been born in the seaside town of Ramsgate and grew up in a prestigious public school called Chatham House, where his father was headmaster. Whitehead had been a stellar student at Chatham, finishing top of the school in English composition, and he went on to attend Lincoln College at Oxford, where he developed a reputation for sociability and kindness that would last the rest of his days. He became a great devotee of the intellectual tavern life: sitting with a handful of friends over dinner, savoring a pipe, telling stories or debating politics or discussing moral philosophy in the late hours of the night. When asked about his college years, Whitehead liked to say that he got more good out of men than he got out of books.

By the time he left Oxford, Whitehead had decided to enter the Anglican Church, and was ordained in London several years later. His religious calling did nothing to abate his fondness for London's taverns, and he frequented the old establishments around Fleet Street—The Cock, The Cheshire Cheese, The Rainbow. Whitehead was liberal in his political views but, as friends often remarked, conservative in his morals. In addition to his religious training, he had a sharp, empirical mind and a good memory for detail. He was also unusually tolerant of maverick ideas, and immune to the bromides of popular opinion. He was often heard saying to friends, "Mind you, the man who is in the minority of one is almost sure to be in the right."

In 1851, the vicar of St. Luke's offered him a position, telling Whitehead that the parish was a place for those who "care more for the approval than the applause of men." At St. Luke's he worked as a kind of missionary to the slum dwellers of Berwick Street, and was a well-regarded and familiar figure in the tumultuous neighborhood. One of Whitehead's contemporaries captured the chaotic sights and sounds of the streets around St. Luke's in that period:

> One does not realize as one passes down Regent Street, how small a distance of street and alley separates "the unknown little from the unknowing great." But to the person who will dive down such entrance to the unknown land of slums of Soho as Beak Street or Berwick Street provides, there is much that will astonish and interest him, if he is a student of the ways of the poor in London. Your cab is suddenly brought up sharp by a coster's barrow, and you are asked if you are going down to St. Luke's. Berwick Street: if you in-

> timate that this is your destination, you are told politely, but with proper Soho emphasis, that you will get through by the end of next week, and you are soon obliged to believe there is truth in the prophecy. Closely ranged side by side in the narrow street are the vendors' stalls and barrows. The cats'-meat man, the fish salesman, the butcher, the fruiterer, the toy-seller, the old rang-and-bone men, jostle and cry their wares. "Prime meat! meat! meat! buy! buy! buy! Here! here! here! veal! veal! fresh-veal today! what's your fancy! Sold, sold again! fish for nothing! cherries ripe!" Your aim is St. Luke's, Berwick Street: you soon see its dim row of dingy semi-domestic, semi-gothic windows. A man is standing just opposite the barred gate skinning eels; you hear a scream, and you know that a poor creature who objects to its fate has slipped from his hand, and is making its way among the crowd.

In the heat and humidity of late August, the smells of Soho would have been unavoidable, wafting up from the cesspools and sewers, from the factories and furnaces. Part of the stench derived from the omnipresence of livestock in the city center. A modern-day visitor time-traveling back to Victorian London wouldn't be surprised to see horses (and, consequently, their manure) in great numbers in the city streets, but he would probably be startled to discover how many farm animals lived in densely packed neighborhoods like Golden Square. Veritable herds would stream through the city; the main livestock market at Smithfield would regularly sell 30,000 sheep in two days' time. A slaughterhouse at the edge of Soho, on Marshall Street, killed an average of five oxen and seven sheep per day, the blood and filth from the animals draining into gulley holes on the street. Without proper barns, residents converted traditional dwellings into "cow houses"—herding twenty-five or thirty cows into a single room. In some cases, cows were lifted into attics via windlass, and shuttered there in the dark until their milk gave out.

Even the pets could be overwhelming. One man who lived on the upper floor at 38 Silver Street kept twenty-seven dogs in a single room. He would leave what must have been a prodigious output of canine excrement to bake in the brutal summer sun on the roof of the house. A charwoman down the street kept seventeen dogs, cats, and rabbits in her single-room flat.

The human crowding was almost as oppressive. Whitehead liked to tell the story of visiting one densely packed household, and asking an impoverished woman there how she managed to get along in such close quarters. "Well, sir," she replied, "we was comfortable enough till the gentleman come in the middle." She then pointed to a chalk circle in the center of the room, defining the region that the "gentleman" was allowed to occupy.

Henry Whitehead's journey that morning would have been a meandering, sociable one: stopping by a coffeehouse largely patronized by machinists, visiting with parishioners in their homes, spending a few minutes down the street from his church with the inmates at the St. James Workhouse, where five hundred of London's impoverished citizens were housed and forced to perform arduous labor through the day. He might have paid a call on the Eley Brothers factory, home to 150 employees churning out one of the most important military inventions of the century: the "percussion cap," which had enabled firearms to be operated in any weather. (Older, flint-based systems were easily disabled by a mild rainshower.) With the outbreak of the Crimean War several months earlier, the Eley brothers were doing a brisk business.

At the Lion Brewery on Broad Street, the seventy workers employed there went about their daily labor, sipping on the malt liquor supplied as part of their wages. A tailor living above the Lewis family at 40 Broad—we know him only as Mr. G—worked his trade, assisted occasionally by his wife. On the sidewalks, the upper echelons of London's street laborers swarmed: the menders and makers, the costermongers and street sellers, hawking everything from crumpets to almanacs to snuff boxes to live squirrels. Henry Whitehead would have known many of these people by name, and his day would have been a steady, comforting stream of sidewalk and parlor conversation. No doubt the heat would have been a primary topic of conversation: the temperature had peaked in the nineties for several straight days, and the city had seen scarcely a drop of rain since the middle of August. There was news from the Crimean War to discuss, as well as the appointment of a new head of the Board of Health, a man by the name of Benjamin Hall, who had vowed to continue the bold sanitation campaign of his predecessor, Edwin Chadwick, but without alienating quite as many people. The city was just finishing Dickens' screed against the industrial coketowns of the north country, *Hard Times,* the final installment of which had run in *Household Words* a few weeks before. And then there were the personal details of daily life—an upcoming marriage, a lost job, a grandchild on the way—which Whitehead would have readily discussed, knowing his parishioners as well as he did. But of all the conversations he had over the first three days of that fateful week, Whitehead would later recall one ironic omission: not one of those conversations broached the topic of cholera.

Imagine an aerial view of Broad Street that week, accelerated in the fashion of a time-lapse movie. Most of the activity would be a blur of urban tumult: "the noisy and the eager, and the arrogant and the froward and the vain . . . [making] their usual uproar," as Dickens put it at the end of *Little Dorrit*. But in all that turbulence, certain patterns appear, like eddies in an otherwise chaotic flow. The streets flex with the Victorian equivalent of rush hour, rising at daybreak and then subsiding with nightfall; streams of people pour into each daily service at St. Luke's; small queues form around the busiest street vendors. In front of 40 Broad Street, as baby Lewis suffers only a few yards away, a single point on the sidewalk attracts a constant—and constantly changing—cluster of visitors throughout the day, like a vortex of molecules winding down a drain.

They are there for the water.

THE BROAD STREET PUMP HAD LONG ENJOYED A REPUTATION as a reliable source of clean well water. It extended twenty-five feet below the surface of the street, reaching down past the ten feet of accumulated rubbish and debris that artificially elevated most of London, through a bed of gravel that stretched all the way to Hyde Park, down to the veins of sand and clay saturated with groundwater. Many Soho residents who lived closer to other pumps—one on Rupert Street and another on Little Marlborough—opted to walk an extra few blocks for the refreshing taste of Broad Street's water. It was colder than the water found at the rival pumps; it had a pleasant hint of carbonation. For these reasons, the Broad Street water insinuated itself into a complex web of local drinking habits. The coffeehouse down the street brewed its coffee with pump water; many little shops in the neighborhood sold a confection they called "sherbet," a mixture of effervescent powder with Broad Street water. The pubs of Golden Square diluted their spirits with pump water.

Even émigrés from Golden Square retained their taste for the Broad Street well. Susannah Eley, whose husband had founded the percussion-cap factory on Broad Street, moved to Hampstead after being widowed. But her sons would regularly fill a jug with Broad

Street water and deliver it to her via cart. The Eley brothers also maintained two large tubs of well water for their employees to enjoy during the workday. With temperatures reaching the mid-eighties in the shade on those late-August days, and no wind to freshen the air, the collective thirst for cool well water must have been intense.

We know a remarkable amount about the quotidian drinking habits of the Golden Square neighborhood on those oppressive days of August 1854. We know that the Eley brothers dispatched a bottle to their mother on Monday, and that she shared it with her visiting niece later that week. We know that a young man visiting his chemist father enjoyed a glass of pump water with his pudding at a restaurant on Wardour Street. We know of an army officer who visited a friend on Wardour Street for dinner and drank a glass of Broad Street water with his meal. We know that the tailor Mr. G sent his wife several times to grab a pitcher of water from the pump outside his workplace.

We also know of the holdouts who did not drink water from the pump that week, for a variety of reasons: the laborers at the Lion Brewery who had their malt liquor supplemented by water supplied by the popular New River Company; a family who normally relied on their ten-year-old girl to fetch water from the pump went dry for a few days as the little girl recovered in bed from a cold. A regular pump-water drinker—and noted ornithologist—named John Gould had declined a glass on that Saturday, complaining that it had a repulsive smell. Despite living a few feet from the pump, Thomas Lewis had never favored its water.

There is something remarkable about the minutiae of all these ordinary lives in a seemingly ordinary week persisting in the human record for almost two centuries. When that chemist's son spooned out his sweet pudding, he couldn't possibly have imagined that the details of his meal would be a matter of interest to anyone else in Victorian London, much less citizens of the twenty-first century. This is one of the ways that disease, and particularly epidemic disease, plays havoc with traditional histories. Most world-historic events—great military battles, political revolutions—are self-consciously historic to the participants living through them. They act knowing that their decisions will be chronicled and dissected for decades or centuries to come. But epidemics create a kind of history from below: they can be world-changing, but the participants are almost inevitably ordinary folk, following their established routines, not thinking for a second about how their actions will be recorded for posterity. And of course, if they do recognize that they are living through a historical crisis, it's often too late—because, like it or not, the primary way that ordinary people create this distinct genre of history is by dying.

Yet something has been lost in the record as well, something more intimate and experiential than stories of pudding and malt liquor—namely, what it *felt like* to contract cholera in that teeming, fraught city, at a time when so little was understood about the disease. We have remarkably detailed accounts of the movements of dozens of individuals during that late-summer week; we have charts and tables of lives and deaths. But if we want to re-create the inner experience of the outbreak—the physical and emotional torment involved—the historical record comes up wanting. We have to use our imaginations.

Sometime on Wednesday, it's likely that the tailor at 40 Broad, Mr. G, began to feel an odd sense of unease, accompanied by a slightly upset stomach. The initial symptoms themselves would be entirely indistinguishable from a mild case of food poisoning. But layered over those physical symptoms would be a deeper sense of foreboding. Imagine if every time you experienced a slight upset stomach you knew that there was an entirely reasonable chance you'd be dead in forty-eight hours. Remember, too, that the diet and sanitary conditions of the day—no refrigeration; impure water supplies; excessive consumption of beer, spirits, and coffee—created a breeding ground for digestive ailments, even when they didn't lead to cholera. Imagine living with that sword of Damocles hovering above your head—every stomach pain or watery stool a potential harbinger of imminent doom.

City dwellers had lived with fear before, and London, of course, had not forgotten its Great Plague and its Great Fire. But for Londoners, the specific menace of cholera was a product of the Industrial Age and its global shipping networks: no known case of cholera on British soil exists before 1831. Yet the disease itself was an ancient one. Sanskrit writings from around 500 B.C. describe a lethal illness that kills by draining water from its victims. Hippocrates prescribed white hellebore blooms as a treatment. But the disease remained largely within the confines of India and the Asian Subcontinent for at least two thousand years. Londoners first took notice of cholera when an outbreak among British soldiers stationed in Ganjam, India, sickened more than five hundred men in 1781. Two years later, word appeared in the British papers of a terrible outbreak that had killed 20,000 pilgrims at Haridwar. In 1817, the cholera "burst forth . . . with extraordinary malignity," as the *Times* reported, tracking through Turkey and Persia all the way to Singapore and Japan, even spreading as far as the Americas until largely dissipating in 1820. England itself was spared, which led the pundits of the day to trot out an entire military parade of racist clichés about the superiority of the British way of life.

But this was merely cholera's shot across the bow. In 1829, the disease began to spread in earnest, sweeping through Asia, Russia, even the United States. In the summer of 1831, an outbreak tore through a handful of ships harbored in the river Medway, about thirty miles from London. Cases inland didn't appear until October of that year, in the northeast town of Sunderland, beginning with a William Sproat, the first Englishman to perish of cholera on his home soil. On February 8 of the following year, a Londoner named John James became the first to die in the city. By outbreak's end, in 1833, the dead in England and Wales would number above 20,000. After that first explosion, the disease flared up every few years, dispatching a few hundred souls to an early grave, and then going underground again. But the long-term trend was not an encouraging one. The epidemic of 1848–1849 would consume 50,000 lives in England and Wales.

All that history would have weighed like a nightmare on Mr. G, as his condition worsened on Thursday. He may have begun vomiting during the night and most likely experienced muscle spasms and sharp abdominal pains. At a certain point, he would have been overtaken by a crushing thirst. But the experience was largely dominated by one hideous process: vast quantities of water being evacuated from his bowels, strangely absent of smell and color, harboring only tiny white particles. Clinicians of the day dubbed this "rice-water stool." Once you began emitting rice-water stools, odds were you'd be dead in a matter of hours.

Mr. G would have been terribly aware of his fate, even as he battled the physical agony of the disease. One of cholera's distinctive curses is that its sufferers remain mentally alert until the very last stages of the disease, fully conscious both of the pain that the disease has brought them and the sudden, shocking contraction of their life expectancy. The *Times* had described this horrifying condition several years before in a long feature on the disease: "While the mechanism of life is suddenly arrested, the body emptied by a few rapid gushes of its serum, and reduced to a damp, dead . . . mass, the mind within remains untouched and clear,—shining strangely through the glazed eyes, with light unquenched and vivid,—a spirit, looking out in terror from a corpse."

By Friday, Mr. G's pulse would have been barely detectable, and a rough mask of blue, leathery skin would have covered his face. His condition would have matched this description of William Sproat from 1831: "countenance quite shrunk, eyes sunk, lips dark blue, as well as the skin of the lower extremities; the nails . . . livid."

Most of this is, to a certain extent, conjecture. But one thing we know for certain: at one p.m. on Friday, as baby Lewis suffered quietly in the room next door, Mr. G's heart stopped beating, barely twenty-four hours after showing the first symptoms of cholera. Within a few hours, another dozen Soho residents were dead.

THERE IS NO DIRECT MEDICAL ACCOUNT OF IT, BUT WITH the hindsight of a century and a half of scientific research, we can describe with precision the cellular events that transformed Mr. G from a healthy, functioning human being to a shrunken, blue-skinned cadaver in a matter of days. Cholera is a species of bacterium, a microscopic organism that consists of a single cell harboring strands of DNA. Lacking the organelles and cell nuclei of the eukaryotic cells of plants and animals, bacteria are, nevertheless, more complex than viruses, which are essentially naked strands of genetic code, incapable of surviving and replicating without having host organisms to infect. In terms of sheer numbers, bacteria are by far the most successful organisms on the planet. A square centimeter of your skin contains most likely around 100,000 separate bacterial cells; a bucket of topsoil would contain billions and billions. Some experts believe that despite their minuscule size (roughly one-millionth of a meter long), the domain of bacteria may be the largest form of life in terms of biomass.

More impressive than their sheer number, though, is the diversity of bacterial lifestyles. All organisms based on the complex eukaryotic cell (plants, animals, fungi) survive thanks to one of two basic metabolic strategies: photosynthesis and aerobic respiration. There may be astonishing diversity in the world of multicellular life—whales and black widows and giant redwoods—but beneath all that diversity lie two fundamental options for staying alive: breathing air and capturing sunlight. The bacteria, on the other hand, make a living for themselves in a dazzling variety of ways: they consume nitrogen right out of the air, extract energy from sulfur, thrive in the boiling water of deep-sea volcanoes, live by the millions in a single human colon (as *Escherichia coli* do). Without the metabolic innovations pioneered by bacteria, we would literally have no air to breathe. With the exception of a few unusual compounds (among them snake venom), bacteria can process all the molecules of life, making bacteria both an essential energy provider for the planet *and* its primary recycler. As Stephen Jay Gould argued in his book *Full House,* it makes for good museum copy to talk about an Age of Dinosaurs or an Age of Man, but in reality it's been one long Age of Bacteria on this planet since the days of the primordial soup. The rest of us are mere afterthoughts.

THE TECHNICAL NAME FOR THE CHOLERA BACTERIUM IS *Vibrio cholerae.* Viewed through an electron microscope, the bacterium looks somewhat like a swimming peanut—a curved rod with a thin, rotating tail called the flagellum that propels the organism, not unlike the outboard motor of a speedboat. On its own, a single *V. cholerae* bacterium is harmless to humans. You need somewhere between 1 million and 100 million organisms, depending on the acidity of your stomach, to contract the disease. Because our minds have a difficult time grasping the scale of life in the microcosmos of bacterial existence, 100 million microbes sounds, intuitively, like a quantity that would be difficult to ingest accidentally. But it takes about 10 million bacteria per milliliter of water for the organism's presence to be at all detectable to the human eye. (A milliliter is roughly 0.4 percent—four thousandths—of 1 cup.) A glass of water could easily contain 200 million *V. cholerae* without the slightest hint of cloudiness.

For those bacteria to pose any threat, you need to ingest the little creatures: simple physical contact can't get you sick. *V. cholerae* needs to find its way into your small intestine. At that point, it launches a two-pronged attack. First, a protein called TCP pili helps the bacteria reproduce at an astonishing clip, cementing the organisms into a dense mat, made up of hundreds of layers, that covers the surface of the intestine. In this rapid population explosion, the bacteria inject a toxin into the intestinal cells. The cholera toxin ultimately disrupts one of the small intestine's primary metabolic roles, which is to maintain the body's overall water balance. The walls of the small intestine are lined with two types of cells: cells that absorb water and pass it on to the rest of the body, and cells that secrete water that ultimately gets flushed out as waste. In a healthy, hydrated body, the small intestine absorbs more water than it secretes, but an invasion of *V. cholerae* reverses that balance: the cholera toxin tricks the cells into expelling water at a prodigious rate, so much so that in extreme cases people have been known to lose up to thirty percent of body weight in a matter of hours. (Some say that the name *cholera* itself derives from the Greek word for "roof gutter," invoking the torrents of water that flow out after a rainstorm.) The expelled fluids contain flakes from the epithelial cells of the small intestine (the white particles that inspired the "rice water" description). They also contain a massive quantity of *V. cholerae.* An attack of cholera can result in the expulsion of up to twenty liters of fluid, with a per milliliter concentration of *V. cholerae* of about a hundred million.

In other words, an accidental ingestion of a million *Vibrio cholerae* can produce a trillion new bacteria over the course of three or four days. The organism effectively converts the human body into a factory for multiplying itself a millionfold. And if the factory doesn't survive longer than a few days, so be it. There's usually another one nearby to colonize.

THE ACTUAL CAUSE OF DEATH WITH CHOLERA IS DIFFICULT to pinpoint; the human body's dependence on water is so profound that almost all the major systems begin to fail when so much fluid is evacuated in such a short period of time. Dying of dehydration is, in a sense, an abomination against the very origins of life on earth. Our ancestors evolved first in the oceans of the young planet, and while some organisms managed to adapt to life on the land, our bodies retain a genetic memory of their watery origin. Fertilization for all animals takes place in some form of water; embryos float in the womb; human blood has almost the same concentration of salts as seawater. "Those animal species that fully adapted to the land did so through the trick of taking their former environment with them," the evolutionary biologist Lynn Margulis writes. "No animal has ever really completely left the watery microcosm. . . . No matter how high and dry the mountain top, no matter how secluded and modern the retreat, we sweat and cry what is basically seawater."

The first significant effect of serious dehydration is a reduction in the volume of blood circulating through the body, the blood growing increasingly concentrated as it is deprived of water. The lowered volume causes the heart to pump faster to maintain blood pressure and keep vital organs—the brain and the kidneys—functional. In this internal triage, nonvital organs such as the gallbladder and spleen begin to shut down. Blood vessels in the extremities constrict, creating a persistent tingling sensation. Because the brain continues to receive a sufficient supply of blood in this early stage, the cholera victim retains a sharp awareness of the attack that *V. cholerae* has launched against his body.

Eventually, the heart fails in its ability to maintain adequate blood pressure, and hypotension sets in. The heart pumps at a frenetic rate, while the kidneys struggle to conserve as much fluid as possible. The mind grows hazy; some sufferers become lightheaded or even pass out. The terrible evacuations of rice-water stools continue. By now, the cholera victim may have lost more than ten percent of his body weight in a matter of twenty-four hours. As the kidneys finally start to fail, the bloodstream re-creates on a much smaller scale the crisis of waste management that helped cholera thrive in so many large cities: waste products accumulate in the blood, fostering a condition called uremia. The victim slips into unconsciousness, or even a coma; the vital organs start to shut down. Within a matter of hours, the victim is dead.

But all around him, in his soaked sheets, in the buckets of rice water at his bedside, in the cesspools and sewers, are new forms of life—trillions of them, waiting patiently for another host to infect.

WE SOMETIMES TALK ABOUT ORGANISMS "DESIRING" CERTAIN environments, even though the organism itself surely has no self-awareness, no feeling of desire in the human sense of the word. Desire in this case is a matter of ends, not means: the organisms wants a certain environment because the setting allows it to reproduce more effectively than other environments: a brine shrimp *desires* salty water, a termite *desires* rotting wood. Put the organism in its desired environment, and the world will have more of that particular creature; take it out, and the world will have less.

In this sense, what the *Vibrio cholerae* bacterium desires, more than anything, is an environment in which human beings have a regular habit of eating other people's excrement. *V. cholerae* cannot be transmitted through the air or even through the exchange of most bodily fluids. The ultimate route of transmission is almost invariably the same: an infected person emits the bacteria during one of the violent bouts of diarrhea that are the disease's trademark, and another person somehow ingests some of the bacteria, usually through drinking contaminated water. Drop it into a setting where excrement eating is a common practice, and cholera will thrive—hijacking intestine after intestine to manufacture more bacteria.

For most of the history of *Homo sapiens,* this dependence on excrement eating meant that the cholera bacterium didn't travel well. Since the dawn of civilization, human culture has demonstrated a remarkable knack for diversity, but eating other humans' waste is as close to a universal taboo as any in the book. And so, without a widespread practice of consuming other people's waste, cholera stayed close to its original home in the brackish waters of the Ganges delta, surviving on a diet of plankton.

In practice, it's not impossible for physical contact with a cholera victim to transmit the disease, but the chance of transmission is slight. In handling soiled linens, for instance, an invisible collection of *V. cholerae* might cluster on a fingertip, where, left unwashed, they might find their way into your mouth during a meal, and shortly thereafter begin their deadly multiplication in your small intestine. From the cholera's point of view, however, this is generally an inefficient way to reproduce: only a small number of people are likely to touch the immediate waste products of another human, particularly one suffering from such a violent and deadly illness. And even if a few lucky bacteria do manage to attach themselves to an errant finger, there's no guarantee that they'll survive long enough to make it to the small intestine.

For thousands of years, cholera was largely kept in check by these two factors: humans on the whole were disinclined to knowingly consume each other's excrement; *and,* on those rare occasions when they did accidentally ingest human waste, the cycle wasn't likely to happen again, thus keeping the bacteria from finding a tipping point where it spread at ever-increasing rates through the population, the way more easily transmitted diseases, like influenza or smallpox, famously do.

But then, after countless years fighting to survive through the few transmission routes available, *V. cholerae* got a lucky break. Humans began gathering in urban areas with population densities that exceeded anything in the historical record: fifty people crammed into a four-story townhouse, four hundred to an acre. Cities became overwhelmed with their human filth. And those very cities were increasingly connected by the shipping routes of the grand empires and corporations of the day. When Prince Albert first announced his idea for a Great Exhibition, his speech included these utopian lines: "We are living at a period of most wonderful transition, which tends rapidly to accomplish that great era to which, indeed, all history points: the realisation of the unity of mankind." Mankind was no doubt becoming more unified, but the results were often far from wonderful. The sanitary conditions of Delhi could directly affect the conditions of London and Paris. It wasn't just mankind that was being unified; it was also mankind's small intestine.

Inevitably, in these sprawling new metropolitan spaces, with their global networks of commerce, lines were crossed: drinking water became laced with sewage. Ingesting small particles of human waste went from being an anomaly to a staple of everyday life. This was good news for *V. cholerae.*

The contamination of drinking water in dense urban settlements did not merely affect the number of *V. cholerae* circulating through the small intestines of mankind. It also greatly increased the lethality of the bacteria. This is an evolutionary principle that has long been observed in populations of disease-spreading microbes. Bacteria and viruses evolve at much faster rates than humans do, for several reasons. For one, bacterial life cycles are incredibly fast: a single bacterium can produce a million offspring in a matter of hours. Each new generation opens up new possibilities for genetic innovation, either by new combinations of existing genes or by random mutations. Human genetic change is several orders of magnitude slower; we have to go through a whole fifteen-year process of maturation before we can even think about passing our genes to a new generation.

The bacteria have another weapon in their arsenal. They are not limited to passing on their genes in the controlled, linear fashion that all multicellular organisms do. It's much more of a free-for-all with the microbes. A random sequence of DNA can float into a neighboring bacterial cell and be immediately enlisted in some crucial new function. We're so accustomed to the vertical transmission of

DNA from parent to child that the whole idea of borrowing small bits of code seems preposterous, but that is simply the bias of our eukaryotic existence. In the invisible kingdom of viruses and bacteria, genes move in a far more indiscriminate fashion, creating many disastrous new combinations, of course, but also spreading innovative strategies at a much faster clip. As Lynn Margulis writes: "All the world's bacteria essentially have access to a single gene pool and hence to the adaptive mechanisms of the entire bacterial kingdom. The speed of recombination over that of mutation is superior: it could take eukaryotic organisms a million years to adjust to a change on a worldwide scale that bacteria can accommodate in a few years."

Bacteria like *Vibrio cholerae,* then, are eminently capable of evolving rapid new characteristics in response to changes in their environment—particularly a change that makes it significantly easier for them to reproduce themselves. Normally, an organism like *V. cholerae* faces a difficult cost-benefit analysis: a particularly lethal strain can make untold billions of copies of itself in a matter of hours, but that reproductive success usually kills off the human body that made that reproduction possible. If those billion copies don't find their way into another intestinal tract quickly, the whole process is for naught; the genes for increased lethality are unable to make new copies of themselves. In environments where the risk of transmission is low, the better strategy is to pursue a low-intensity attack on the human host: reproduce in smaller numbers, and keep the human alive longer, in hopes that over time some bacterial cells will find their way to another intestine, where the process can start all over again.

But a dense urban settlement with contaminated water supplies eliminates *V. cholerae*'s dilemma. There's no incentive not to reproduce as violently as possible—and thus kill your host as quickly as possible—because there's every likelihood that the evacuations from the current host will be swiftly routed into the intestinal tract of a new one. The bacterium can invest all its energy in sheer reproductive volume, and forget about longevity.

It goes without saying that the bacteria are not in any way conscious of developing this strategy. The strategy evolves on its own, as the overall population balance of *V. cholerae* changes. In a low-transmission environment, lethal strains die out, and mild ones come to dominate the population. In high-transmission environments, the lethal strains quickly outnumber the mild ones. No single bacterium is aware of the cost-benefit analysis, but thanks to their amazing capacity for adaptation, they're able to make the analysis as a group, each isolated life and death serving a kind of vote in a distributed microbial assembly. There is no consciousness in the lowly bacterium. But there is a kind of group intelligence nonetheless.

Besides, even human consciousness has its limits. It tends to be very acute on the scales of human existence, but as ignorant as the bacteria on other scales. When the citizens of London and other great cities first began gathering together in such extraordinary number, when they began building elaborate mechanisms for storing and removing their waste, and pulling drinking water from their rivers, they did so with conscious awareness of their actions, with some clear strategy in mind. But they were entirely unaware of the impact that those decisions would have among the microbes: not just in making the bacteria more numerous, but also in transforming their very genetic code. The Londoner enjoying his new water closet or his expensive private water supply from the Southwark Water Company was not just engineering his private life to make it more convenient and luxurious. He was also, unwittingly, reengineering the DNA of *V. cholerae* with his actions. He was making it into a more efficient killer.

THE TRAGIC IRONY OF CHOLERA IS THAT THE DISEASE HAS A shockingly sensible and low-tech cure: water. Cholera victims who are given water and electrolytes via intravenous and oral therapies reliably survive the illness, to the point where numerous studies have deliberately infected volunteers with the disease to study its effects, knowing that the rehydration program will transform the disease into merely an uncomfortable bout of diarrhea. You would think that the water cure might have occurred to some of the physicians of the day: the ill were discharging prodigious amounts of water, after all. If you were looking for a cure, wouldn't it be logical to start with restoring some of those lost fluids? And indeed, one British doctor, Thomas Latta, hit upon this precise cure in 1832, months after the first outbreak, injecting salty water into the veins of the victims. Latta's approach differed from the modern treatment only in terms of quantity: liters of water are necessary to ensure a full recovery.

Tragically, Latta's insight was lost in the swarming mass of cholera cures that emerged in the subsequent decades. Despite all the technological advances of the Industrial Age, Victorian medicine was hardly a triumph of the scientific method. Reading through the newspapers and medical journals of the day, what stands out is not just the breadth of remedies proposed, but the breadth of people involved in the discussion: surgeons, nurses, patent medicine quacks, public-health authorities, armchair chemists, all writing the *Times* and the *Globe* (or buying classified advertising there) with news of the dependable cure they had concocted.

Those endless notices reflect a strange historical overlap, one we have largely outgrown—the period *after* the rise of mass communications but *before* the emergence of a specialized medical science.

Ordinary people had long cultivated their folk remedies and homespun diagnoses, but until newspapers came along, they didn't have a forum beyond word of mouth to share their discoveries. At the same time, the medical division of labor that we now largely take for granted—researchers analyze diseases and potential cures, doctors prescribe those cures based on their best assessment of the research—had only reached an embryonic state in the Victorian age. There was a growing medical establishment—best embodied by the prominent journal *The Lancet*—but its authority was hardly supreme. You didn't need an academic degree to share your cure for rheumatism or thyroid cancer with the world. For the most part, this meant that the newspapers of the day were filled with sometimes comic, and almost always useless, promises of easy cures for diseases that proved to be far more intractable than the quacks suggested. But that anarchic system also made it possible for genuine visionaries to route around the establishment, particularly when the establishment had its scientific head in the sand.

The prominence of quack cures also had an unexpected side effect: it helped create an entire rhetoric of advertising—as well as a business model for newspapers and magazines—that has lasted for more than a century. By the end of the 1800s, patent-medicine manufacturers were the leading advertisers in the newspaper business, and as the historian Tom Standage observes, they were "among the first to recognize the importance of trademarks and advertising, of slogans, logos. . . . Since the remedies themselves usually cost very little to make, it made sense to spend money on marketing." It has become a cliché to say that we now live in a society where image is valued over substance, where our desires are continually stoked by the illusory fuel of marketing messages. In a real sense that condition dates back to those now quaint notices running in the columns of

Victorian newspapers, promising an endless litany of cures bottled in one marvelously inexpensive elixir.

Not surprisingly, the patent-medicine industry was eager to provide a cure for the most menacing disease of the nineteenth century. A naïve reader of the London *Times* classifieds in August of 1854 might have naturally assumed that the cholera was on its way out, given all the cures that seemed readily available:

> FEVER and CHOLERA.—The air of every sick room should be purified by using SAUNDER'S ANTI-MEPHITIC FLUID. This powerful disinfectant destroys foul smells in a moment, and impregnates the air with a refreshing fragrance. —J.T. Saunders, perfumer, 316B, Oxford-street, Regent-circus; and all druggists and perfumers. Price 1s.

As laughable as the patent-medicine adverts seem to us today, they nonetheless provoked irate letters complaining about the injustice of keeping these expensive cures out of reach of the lower classes:

> Sir,—I have observed lately several letters in your influential journal, treating upon the present much-talked-of subject—the enormous price of castor oil as retailed by the druggists. . . . One man in this town [has] boldly come forward and made a public announcement, in the shape of placards upon the walls, that he is prepared to sell the finest cold-drawn castor oil at 1d. per ounce, and it is to be hoped that his example will be universally followed. Sure, Sir, when a druggist himself is candid enough to publish to the world that he can afford to sell this article at 1d. per ounce instead of 3d. and by so doing have a sufficient profit thereby, can there now be any doubt whatever in the minds of the people that this class of tradesman have for

> many years past been reaping a great harvest by retailing castor oil to the poor at such immense gains.

You can see in these sentences the beginning of another modern sensibility: the outrage that is now directed against the price gouging of multinational drug companies. But at least Big Pharma is, more often than not, selling something that actually works. It is hard to say which would be a worse offense: selling castor oil with such high profit margins, or giving it away as a charitable act. At least the high prices discouraged people from employing the noxious stuff.

One step up the food chain were the letters to the *Times,* often written by accredited medical men, offering up their remedy (or disputing another's) for less obviously commercial ends. In the late summer of 1854, the surgeon-in-chief of the city police, G. B. Childs, had taken to writing the *Times* with descriptions of his fail-safe remedy for cholera's most telltale symptom: diarrhea. This is his letter from the eighteenth of August:

> Will you . . . kindly allow me a space in your columns, not only to reiterate what I have already with reference to ether and laudanum, but to explain how, in my opinion, these remedies act when taken into the stomach? If any corroborative testimony of its efficacy be further required, I would ask those who might be skeptical of its merits to call at any one of the police stations in the city of London, where a supply of the medicine is kept and satisfy themselves of the estimation in which it is held by the members of the force. . . . You want something which will act immediately without requiring the slow, and in these cases uncertain, process of digestion. If the properties of opium are valuable, and they are acknowledged to be such

by all authorities, the sooner these properties are brought into active operation the better. . . . In conclusion, Sir, I beg to observe that in submitting these remedies to your numerous readers I feel that, as a public officer, I am only discharging a public duty.

Formally, those closing solemn statements are typical of the genre, and of course their solemnity plays against the modern reader's amusement at the remedy itself. After all, we have here a chief law enforcement official writing into the daily paper essentially to encourage people to ingest heroin to treat their upset stomachs—and if the readers don't believe him, they should head down to the nearby squad house to hear firsthand how highly regarded the "medicine" is by the police force. Not exactly a "war on drugs" sentiment, although not entirely without merit medically: constipation is a reliable side effect of opiate abuse.

Cholera remedies were a running dialogue in the papers of the day, a source of endless debate. One M.D. would write in endorsing his cocktail of linseed oil and hot compresses on Tuesday, and by Thursday another would be running off a list of patients who had died after following precisely such a treatment.

Sir,—Induced by the very favourable results of the use of castor oil in cholera, as reported by Dr. Johnson, I have just put his practice to the test of experience, and I regret to say with signal failure. . . .

Sir,—Let me entreat your metropolitan readers not to be led by the letter of your correspondent into the belief that smoke is in any way a preventative of cholera, or can in any degree influence the prevalence of epidemic disease. . . .

The constant squabbling between medical authorities in the papers eventually hit a point of self-parody. The week of the Broad Street outbreak, *Punch* went to press with a lacerating editorial titled "Who Shall Decide When Doctors Disagree?"

> It really is nauseating to witness the quantity of doctor's stuff that is allowed to run down the columns of the newspapers. It will be necessary at last to proceed against the public press as a public nuisance if we have much more of the "foul and offensive matter" circulating under our noses every day at our breakfast tables to an extent highly dangerous to the health, the patience, and the nerves of the reading community. If the doctors who write to the papers would agree in their prescriptions for cholera, the public might feel grateful for the trouble taken, but when one medical man's "infallible medicine" is another man's "deadly poison," and the specific of to-day is denounced as the fatal drug of to-morrow, we are puzzled and alarmed at the risk we run in following the doctors' contradictory directions.

Ordinary doctors possessed no less unanimity in their treatment of cholera than the patent-medicine impresarios or the newspaper letter-writers. Sometimes the cholera was treated with leeches, based on the humoral theory that whatever seemed wrong with the patient should be removed from the patient: if the cholera sufferer's blood was unusually thick, thanks to dehydration, then the patient needed to lose more blood. Contrary to G. B. Childs' advice, many doctors prescribed laxatives to combat a disease that was already expelling fluids from the body at a lethal rate. Purgatives like castor oil or rhubarb were widely prescribed. Physicians were also inclined to recommend brandy as a treatment, despite its known dehydrating effects. While these were not quite examples of the cure being worse than the disease—cholera set the bar quite high, as diseases go—many of the proposed remedies exacerbated the physiological crisis that cholera induced. The few positive effects, such as they were, were mostly placebo in nature. And of course, in this elaborate mix of homespun remedies, commercial elixirs, and pseudoscientific prescriptions you would almost never find the real advice that the patients needed to hear: *rehydrate.*

ON FRIDAY MORNING, THE GROWING SENSE OF DREAD HAD not yet expanded beyond the borders of the Golden Square neighborhood. The heat wave had finally broken, and the rest of the city savored the cool, clear weather. There was no way to know that in their midst a terrible outbreak was claiming its first victims. The *Morning Chronicle*'s one item about cholera sounded an upbeat note, reflecting on its diminished presence on the front lines of the Crimean War: "Having at length emerged from the dangers of the month of August, we may hope to behold the abatement of pestilence at the seat of war, and the resumption of active operations. There seems to be little doubt that cholera has done its worst, and that its ravages in the allied army are very considerably mitigated, both in extent and virulence; and the fleet also, which was attacked somewhat later, appears to have now passed the crisis of the disorder."

But within the crowded parlors of Golden Square, the fear was inescapable. The outbreak hit a new peak a few hours before midnight on Thursday. Hundreds of residents had been seized by the disease within a few hours of one another, in many cases entire families, left to tend for themselves in dark, suffocating rooms.

Those fearful scenes—a family crammed together in a room, suffering through the most excruciating private torture as a group—are perhaps the most haunting of all the images of the Broad Street outbreak. Families continue to perish together in the developed world, of course, but such catastrophes usually unfold over the space of seconds or minutes, in car accidents and plane crashes or natural disasters. But a family dying together, slowly, agonizingly, with full awareness of their fate—that is a supremely dark chapter in the book of death. That it continues on as a regular occurrence in certain parts of the world today should be a scandal to us all.

Overnight, Henry Whitehead's sociable rounds as assistant curate of St. Luke's had become a death vigil. Within a few minutes of dawn, he had been called to a house where four people lay near death, their skin already taut and blue. Each house he visited that morning presented the same horrifying scene: a neighborhood on the edge of oblivion. Just before noon, he ran across the scripture reader and another curate from St. Luke's, and found both men had encountered the same devastation in their passages through the neighborhood.

Whitehead's travels took him to four houses along Peter Street near Green's Court, where he found the disease in full fury. Half the occupants, it seemed, had fallen ill in the past twenty-four hours. In one of the grandest of the homes, standing at the northwest corner of Green's Court, all twelve residents would eventually perish. Yet the cholera had largely spared the cramped and grubby quarters on Green's Court itself. (Only five of the two hundred living there would eventually die.) When Whitehead stopped by one of the filthiest houses in the district, he found, to his amazement, that not one of its inhabitants had fallen ill.

The contrast was striking, especially as the four houses on Peter Street had been commended by the parish authorities for their cleanliness during a 1849 survey of the neighborhood, while the survey had found nothing but squalor and soot in the surrounding houses. It occurred to Whitehead that, contrary to the prevailing wisdom, the sanitary conditions of the homes seemed to have no predictive power where the disease was concerned.

Such observations were characteristic of the young deacon, on a number of levels. There is, first, his composure and probing intelligence in a time of great chaos, but also his willingness to challenge orthodoxy, or at least submit it to empirical scrutiny. That scrutiny itself relied on his firsthand knowledge of the neighborhood and its residents. He detected these early patterns in the disease's course precisely because he possessed such a fine-grained understanding of the environment: the houses that had been praised for their sanitary conditions, and the ones considered to be the filthiest on their blocks. Without that kind of knowledge, the platitudes would have been far easier to settle back on.

There were other medical detectives on the streets of Soho that day, looking for clues, building chains of cause and effect. Minutes before sunrise on Saturday morning, John Rogers, a medical officer based on Dean Street, made his way from Walker's Court to Berwick Street, struggling to schedule visits to all the patients who had fallen ill in the previous twenty-four hours. Rogers had seen cholera outbreaks before, but already it was clear that something exceptional was under way in Golden Square. Cholera rarely exploded through a population; it could kill by the thousands, of course, but the carnage usually took months or years to unfold. Rogers was starting to hear accounts of entire households falling ill overnight. And this strain of the disease seemed to do its damage with a terrifying velocity: sufferers were going from complete health to death in twelve hours.

Rogers' itinerary took him past 6 Berwick Street, home to a well-regarded local surgeon by the name of Harrison whom Rogers knew professionally. As Rogers crossed the front of the house, a powerful stench overtook him and he stumbled on the sidewalk for a few seconds, holding back the urge to vomit. He would later describe it as one of the "most sickening and nauseating odours it has ever been misfortunate to inhale in this metropolis." Once Rogers had composed himself, he stepped back and observed that the smell was coming from a gulley hole by the side of the road, a slit on the edge of the curb designed to capture water runoff during storms. Rogers didn't stay long enough to determine what foul combination of decaying matter lay behind the hole. But he thought to himself as he marched onward that the stench was strong enough to pervade the entire residence at number 6.

A few hours later he learned that the surgeon Harrison had expired that morning. Rogers burst out with an immediate diagnosis: "That gulley hole has destroyed him!" He began fulminating against the dreadful sanitary conditions in the city that had led to the catastrophe around him. But the deaths were just beginning. By the end of the week, seven other residents of 6 Berwick would come down with cholera. All but one would perish.

Back at 40 Broad Street, the Lewis infant had descended into an exhausted silence over the night. In the mid-morning hours, her parents called their Dr. Rogers, who had treated the infant earlier in the week. By the time he arrived, a few minutes past eleven, baby Lewis was dead.

THAT AFTERNOON, WHITEHEAD VISITED A FAMILY OF SIX (call them the Waterstones, since no record of their names exists) with whom he had long enjoyed a close connection: two grown sons and two adolescent girls living with their parents in three connected ground-floor rooms off of Golden Square. When he arrived, he found the younger sister, whose wit and good cheer had always impressed Whitehead, fading in and out of consciousness, after a violent and sleepless night suffering from the disease. She was surrounded by her brothers and by a neighbor who had valiantly dropped in to lend a hand. While Whitehead spoke to the men in hushed tones, huddled together in the small center room of the flat, the girl seemed to regain some of her acuity.

At one point she pulled her head up and asked after her mother and sister. Her brothers fell silent. The girl looked anxiously toward the two closed doors at either side of the room. She knew the truth before a word was spoken: behind each door there lay a coffin. She could hear the weeping of her father, draped over the body of his dead wife in the dark of the shuttered front parlor.

Half the neighborhood, it seemed, had shuttered themselves inside, either to suffer in isolation or to ward off whatever foul effusion had brought the plague to the neighborhood. Outside, in the strangely incongruous glare of a summer afternoon, at the top of Berwick Street, a yellow flag was raised to alert the residents that the cholera had struck. The gesture was superfluous. You could see the dead being wheeled down the street by the cartload.

JOHN SNOW

Sunday, September 3

THE INVESTIGATOR

BY SUNDAY MORNING, A STRANGE QUIET HAD OVERTAKEN the streets of Soho. The usual chaos of the streetsellers had disappeared; most of the neighborhood's residents had either evacuated or were suffering behind their doors. Seventy of them had perished over the preceding twenty-four hours, hundreds more were at the very edge of death. Out in front of 40 Broad, the pump attracted only a handful of stragglers. The most common sight on the streets were the priests and doctors making their frantic rounds.

Word of the outbreak had traveled through the wider city and beyond. The chemist's son who had enjoyed his pudding days before on Wardour Street died on that Sunday at his home in Willesden. The entire city held its breath as it took in refugees from the embattled neighborhood, waiting to see if the outbreak in Golden Square would be re-created on a larger scale in the coming days. Seventy deaths in a single parish was not an uncommon number to hear in an age of cholera epidemics. But it normally took months for the disease to chalk up so many victims. The Broad Street strain of cholera—whatever it was, wherever it had come from—had managed that terrible feat in a single day.

While the disease had remained largely confined to an area of roughly five square blocks, the rest of Soho was on high alert. Many packed their bags and visited friends or family who lived in the country or other parts of the city; some locked the doors and shuttered the windows. The vast majority steered clear of the Golden Square neighborhood at all cost.

But one Soho regular had been following the case closely from his residence at Sackville Street on the southwestern edge of the neighborhood. Sometime near dusk he set out from his home, marching through the empty streets, directly into the heart of the outbreak. When he reached 40 Broad, he stopped and examined the pump for a few minutes in the fading light. He drew a bottle of water from the well, stared at it for a few seconds, then turned and made his way back to Sackville Street.

JOHN SNOW WAS IN HIS FORTY-SECOND YEAR, AND SINCE his early thirties he had by any measure enjoyed a remarkable streak of professional achievement. Unlike most members of the medical establishment or the sanitary reform movement, Snow had been born into a family of modest means, the eldest son of a Yorkshire laborer. A quiet, serious child with intellectual ambitions beyond his humble origins, Snow had apprenticed at the age of fourteen to a surgeon in Newcastle-on-Tyne. At the age of seventeen he read John Frank Newton's influential 1811 manifesto *The Return to Nature: A Defence of the Vegetable Regimen* and promptly converted to vegetarianism. Shortly thereafter, he became a strict teetotaler. He would largely avoid meat and alcohol for the rest of his adult life.

As an apprentice in Newcastle, Snow saw the ravages of cholera firsthand when the disease struck in late 1831. He treated the survivors of a particularly brutal outbreak in a local mine, the Killingworth Colliery. The young Snow observed that the sanitary conditions in the mine were dreadful, with workers granted no separate quarters to relieve themselves, thus forcing them to eat and defecate in the same dark, stifling caverns. The idea that the cholera outbreak was rooted in the social conditions of these impoverished workers—and not in any innate susceptibility to the disease—lodged in the back of Snow's mind as the cholera ran its course. It was only a partially realized thought, nowhere near a genuine theory. But it stayed with him, nonetheless.

A young Englishman interested in the medical life during the first half of the nineteenth century had three primary career paths open to him. He could apprentice with an apothecary and then eventually land a license from the Society of Apothecaries, which would grant him the right to concoct medicines prescribed by physicians. After some training, he would be free to embark on his own practice, treating patients with the woeful remedies of the day, probably dabbling in minor surgery or dentistry on the side. The more ambitious individual would go on to study at a medical school, and later join the Royal College of Surgeons of England, becoming a bona fide general practitioner and surgeon, performing a host of different tasks: everything from treating minor colds to excising bunions to amputating limbs. Beyond that lay the university degree Doctor of Medicine, whose recipients were conventionally called physicians, as opposed to the lower orders of surgeons and apothecaries. A university degree opened doors to the private hospitals, where one could rub shoulders with the wealthy benefactors who endowed them.

Snow realized at an early age that his ambitions extended beyond that of a provincial apothecary. He had moved back to York in 1835 and involved himself in the growing temperance movement there. But at the age of twenty-three, he decided to follow the classic itinerary of the bildungsroman genre that dominated the nineteenth-century novel: a provincial young man with dreams of greatness sets out for the big city to make a name for himself. Snow's journey to London was typical of the earnest young doctor-in-training: he eschewed both horse and carriage and walked a meandering two-hundred-mile route alone.

In London, Snow settled in Soho and enrolled in the Hunterian School of Medicine. Within two years he had received both his apothecary and surgeon's license and established a general practice at 54 Frith Street in London, about a five-minute walk east from Golden Square. Setting up shop as a doctor in those days required an entrepreneurial spirit. The competition was intense among London's new medical middle class—four other surgeons had offices within a few blocks of Snow, though the only physicians nearby resided across Soho in Golden Square. Despite the proximity of so many rivals, Snow quickly established a successful practice. Temperamentally he was not the cliché of the friendly, garrulous general practitioner; his bedside manner was taciturn and emotionally flat. But he was a superb doctor: observant, quick-witted, and possessed of an exceptional memory for past cases. Snow was as free from superstition and dogma as it was possible to be in those days, though he was inevitably limited in his effectiveness by the conceptual dead ends and distortions of early Victorian medicine. The idea of microscopic germs spreading disease would have been about as plausible as the existence of fairies to most practicing doctors of the day. And as Surgeon-in-Chief G. B. Childs' letter-writing campaign to the *Times* suggested, laudanum was regularly prescribed for almost any ailment. The Victorian medical refrain was, essentially: Take a few hits of opium and call me in the morning.

Seemingly bereft of anything resembling a traditional social life, Snow spent his time away from patients working on side projects that grew out of his surgeon's practice but which also suggested the ultimate range of his ambition. He began writing in to the local journals, opining on medical and public-health issues of the day. His first published paper, addressing the use of arsenic in the preservation of cadavers, appeared in *The Lancet* in 1839. He went on to publish nearly fifty articles in the following decade, on a staggering range of subjects: lead poisoning, resuscitating stillborn children, blood vessels, scarlet fever, and smallpox. He wrote in to *The Lancet* with so many critiques of sloppy science that the editor eventually scolded him gently in print, suggesting that "Mr. Snow might better employ himself in producing something, than in criticizing the production of others."

Snow clearly had his mind set on producing something of his very own, and he saw advanced degrees as a crucial bridge to that end. In 1843, he earned his bachelor of medicine degree from the University of London. A year later he had passed the challenging M.D. exams, placing in the first division of students. He was now, officially, Dr. John Snow. By most standards, he was already a remarkable success story: a laborer's son who now had a thriving medical practice and a vibrant career as a researcher and lecturer. At the recommendation of one of his former professors, he had been invited to join the Westminster Medical Society, where he quickly became a respected and active member. Any number of physicians would have settled into that comfortable realm, pursuing only the incremental advances of tending to increasingly well-to-do clientele and elevating their own social prestige in the process. But Snow was oblivious of the trappings of London's polite society; what drove him, more than anything, were problems that needed solutions, filling in the blind spots in the medical establishment's vision of the world.

Snow would continue to work as a practicing physician for the rest of his life, but his eventual fame would come from his pursuits outside the consulting room. Snow did not aim low in his investigations. He would play a defining role in the battle against the era's most relentless killer. But before he could tackle cholera, John Snow set his sights on one of the most excruciating deficiencies of Victorian medicine: pain management.

WHERE SHEER PHYSICAL BRUTALITY WAS CONCERNED, THERE was little in Victorian society that rivaled the professional medical act of surgery. Lacking any form of anesthesia beyond opium or alcohol—both of which could only be applied in moderation, given their side effects—surgical procedures were functionally indistinguishable from the most grievous forms of torture. Surgeons prided themselves on their speed above all else, since extended procedures were unbearable for both doctor and patient. Procedures that would now take hours to complete were executed in three minutes or less, to minimize the agony. One surgeon boasted that he could "amputate a shoulder in the time it took to take a pinch of snuff."

In 1811, the British author—and longtime Soho resident—Fanny Burney underwent a mastectomy in Paris. She described the experience in a letter to her sister a year later. After drinking a wine cordial as her sole form of painkiller, she settled into the ominous closet that had been assembled by the team of seven doctors in her home, lined with compresses and bandages and gruesome surgical tools. She lay down on the makeshift bed, and the doctors covered her face with a light handkerchief. "When the dreadful steel was plunged into the breast, cutting through veins, arteries, flesh, nerves, I needed no injunction not to restrain my cries. I began a scream that lasted unintermittingly during the whole time of the incision, and I almost marvel that it rings not in my ears still! So excruciating was the agony. . . . I then felt the knife rackling against the breastbone, scraping it! This performed, while I remained in utterly speechless torture." Before passing out in near shock after the procedure, she caught a glimpse of her primary doctor—"pale nearly as myself, his face streaked with blood and its expression depicting grief, apprehension, and almost horror."

In October of 1846, at Massachusetts General Hospital in Boston, a dentist named William Morton gave the first public demonstration of the use of ether as an anesthetic. Word quickly spread across the Atlantic, and by mid-December, a London dentist, James Robinson, had begun using ether on his patients, usually with a small audience of amazed medical men in attendance. On December 28, he performed another successful extraction. In the room, watching, with his usual quiet and observant manner, was John Snow.

By the turn of the new year, the excitement over ether had spilled beyond the medical community and into the popular press. *Punch* was running mock editorials advocating its use on difficult wives. But the miracle anesthetic was unreliable in practice. Some applications would work flawlessly: the patient would nod off for the length of the surgery, and then awaken minutes later with no memory of the procedure, and a greatly minimized feeling of pain. But other patients would fail to go under, or return to consciousness abruptly in the middle of a particularly delicate operation. More than a few patients never woke up at all.

Snow quickly hypothesized that the unreliability of ether was likely a problem of dosage, and embarked on a series of interlinked experiments to determine the best mechanism for delivering the miracle gas. From his earlier studies, Snow knew the concentration of any gas varied dramatically with temperature, and yet the early adopters of etherization had failed to take room temperature into account in their procedures. A patient etherized in a chilly room would end up with a significantly lower dose than one etherized in a room warmed by a roaring fire. By mid-January, Snow had compiled a "Table for Calculating the Strength of Ether Vapour." Increasing the temperature by twenty degrees Fahrenheit would nearly double the dosage. The *Medical Times* published Snow's table at the end of January.

While compiling the data for his numerical breakdown of ether's properties, Snow had begun collaborating with a surgeon's instrument maker named Daniel Ferguson in making an inhaler that would allow maximum control of the dosage. Snow's idea was to adapt the well-known Julius Jeffrey vaporizer for the purposes of delivering ether, forcing it through a metal spiral at the center of the device, thus maximizing the surface area of metal exposed to the gas as it traveled to the patient's mouth. The unit would be placed in a vat of heated water that would transmit its warmth to the metal contraption, where it would raise the temperature of the ether. All the doctor needed to control was the temperature of the water; the device would do the rest. Once the doctor had a reliable fix on the ether's temperature, he could determine the proper dose with little variation. Snow first presented his device to the Westminster Society on January 23, 1847.

Snow's productivity during this period is truly astounding, when you think that the very concept of etherization simply hadn't existed three months before. Not only had Snow detected one of the fundamental properties of the gas within two weeks of first seeing it applied, he had also engineered a state-of-the-art medical device to deliver it. And his research had only begun: in the following months, he explored the biology of etherization: everything from the initial intake of the gas in the lungs, to its circulation through the bloodstream, all the way to its psychological effects. When the medical community shifted its focus to the rival anesthetic chloroform later in 1847, Snow immersed himself in its properties as well. By the end of 1848 he had published a seminal monograph on the theory and practice of anesthesia: *On the Inhalation of the Vapour of Ether in Surgical Operations.*

Snow managed to build his mastery of this embryonic field almost entirely through research conducted in his own home. He maintained a small menagerie in his Frith Street quarters—birds, frogs, mice, fish—where he spent countless hours watching the creatures' response to various dosages of ether and chloroform. He also used his medical practice as a source of experimental data, but was not above using himself as a test subject. There is something wonderful—and more than a little ironic—in this image of Snow the teetotaler, arguably the finest medical mind of his generation, performing his research. He sits alone in his cluttered flat, frogs croaking around him, illuminated only by candlelight. After a few minutes tinkering with his latest experimental inhaler, he fastens the mouthpiece over his face and releases the gas. Within seconds, his head hits the desk. Then, minutes later, he wakes, consults his watch through blurred vision. He reaches for his pen, and starts recording the data.

SNOW'S MASTERY OF ETHER AND CHLOROFORM RAISED HIM to a new echelon in the London medical world. He became the most sought-after anesthesiologist in the city, assisting with hundreds of operations a year. By the 1850s, a growing number of doctors were recommending chloroform as a palliative for the discomfort of childbirth. As the birth of her eighth child approached in the spring of 1853, Queen Victoria decided to give chloroform a try, encouraged by the scientifically astute Prince Albert. Her choice of an anesthesiologist was an obvious one. Snow gave the episode a few more words than usual in his casebooks, though his tone did not betray the magnitude of the professional honor that had been bestowed upon him:

> Thursday 7 April: Administered Chloroform to the Queen in her confinement. Slight pains had been experienced since Sunday. Dr. Locock was sent for about nine o'clock this morning, stronger pains having commenced, and he found the os uteri had commenced to dilate a very little. I received a note from Sir James Clark a little after ten asking me to go to the Palace. I remained in an apartment near that of the Queen, along with Sir J. Clark, Dr. Ferguson and (for the most part of the time) Dr. Locock till a little a [*sic*] twelve. At a twenty minutes past twelve by a clock in the Queen's apartment I commenced to give a little chloroform with each pain, by pouring about 15 minims [0.9 ml] by measure on a folded handkerchief. The first stage of labour was nearly over when the chloroform was commenced. Her Majesty expressed great relief from the application, the pains being very trifling during the uterine contractions, and whilst between the periods of contraction there was complete

> ease. The effect of the chloroform was not at any time carried to the extent of quite removing consciousness. Dr. Locock thought that the chloroform prolonged the intervals between the pains, and retarded the labour somewhat. The infant was born at 13 minutes past one by the clock in the room (which was 3 minutes before the right time); consequently the chloroform was inhaled for 53 minutes. The placenta was expelled in a very few minutes, and the Queen appeared very cheerful and well, expressing herself much gratified with the effect of the chloroform.

Snow's research into anesthesia had elevated him from a surgeon of humble origins to the very apogee of Victorian London. But, in a way, the most impressive thing about his research was not the levels of social class that he traversed but rather the intellectual strata, the different scales of experience that his mind crossed so effortlessly. Snow was a truly *consilient* thinker, in the sense of the term as it was originally formulated by the Cambridge philosopher William Whewell in the 1840s (and recently popularized by the Harvard biologist E. O. Wilson). "The Consilience of Inductions," Whewell wrote, "takes place when an Induction, obtained from one class of facts, coincides with an Induction obtained from another different class. Thus Consilience is a test of the truth of the Theory in which it occurs." Snow's work was constantly building bridges between different disciplines, some of which barely existed as functional sciences in his day, using data on one scale of investigation to make predictions about behavior on other scales. In studying ether and chloroform, he had moved from the molecular properties of the gas itself, to its interactions with the cells of the lungs and the bloodstream, to the circulation of those properties through the body's overall system, to the psychological effects produced by these biological changes. He even ventured beyond the natural world into the design of technology that would best reflect our understanding of the anesthetics. Snow was not interested in individual, isolated phenomena; he was interested in chains and networks, in the movement from scale to scale. His mind tripped happily from molecules to cells to brains to machines, and it was precisely that consilient study that helped Snow uncover so much about this nascent field in such a shockingly short amount of time.

And yet, there was a ceiling to his intellectual pursuit of ether and chloroform: his research stopped at the scale of the individual subject. The next step up the chain—the larger, connected world of cities and societies, of groups, not individuals—did not factor into his anesthesia investigations. He might have attended on the queen's body, but the body politic remained outside Snow's frame of reference.

Cholera would change all that.

We don't know exactly what sequence of events turned John Snow's interest toward cholera in the late 1840s. For this working physician and researcher, of course, the disease would have been a constant presence in his life. There may in fact have been a direct link to his practice as an anesthesiologist, since chloroform had been (wrongly) championed as a potential cure for cholera by some early adopters who were less rigorous in their empiricism than Snow. Certainly, the outbreak of 1848–1849, the most severe British outbreak in more than a decade, made cholera one of the most urgent medical riddles of its time. For a man like Snow, obsessed with both the practice of medicine and the intellectual challenge of science, cholera would have been the ultimate quarry.

There were practically as many theories about cholera as there were cases of the disease. But in 1848, the dispute was largely divided between two camps: the contagionists and the miasmatists. Either cholera was some kind of agent that passed from person to person, like the flu, or it somehow lingered in the "miasma" of unsanitary spaces. The contagion theory had attracted some followers when the disease first reached British soil in the early 1830s. "We can only suppose the existence of a poison which progresses independently of the wind, of the soil, of all conditions of the air, and of the barrier of the sea," *The Lancet* editorialized in 1831. "In short, one that makes mankind the chief agent for its dissemination." But most physicians and scientists believed that cholera was disease spread via poisoned atmosphere, not personal contact. One survey of published statements from U.S. physicians during the period found that less than five percent believed the disease was primarily contagious.

By the late 1840s the miasma theory had established a far more prestigious following: the sanitation commissioner, Edwin Chadwick; the city's main demographer, William Farr; along with many other public officials and members of Parliament. Folklore and superstition were also on the side of the miasmatists: the foul inner-city air was widely believed to be the source of most disease. While no clear orthodoxy existed regarding the question of cholera's transmission, the miasma theory had far more adherents than any other explanatory model. Remarkably, in all the discussion of cholera that had percolated through the popular and scientific press since the disease had arrived on British soil in 1832, almost no one suggested that the disease might be transmitted by means of contaminated water. Even the contagionists—who embraced the idea that the disease was transmitted from person to person—failed to see merit in the waterborne scenario.

Snow's detective work into cholera began when he noticed a

telling detail in the published accounts of the 1848 epidemic. Asiatic cholera had been absent from Britain for several years, but it had recently broken out on the Continent, including the city of Hamburg. In September of that year, the German steamer *Elbe* docked in London, having left port at Hamburg a few days earlier. A crewman named John Harnold checked into a lodging house in Horsleydown. On September 22, he came down with cholera and died within a matter of hours. A few days later, a man named Blenkinsopp took over the room; he was seized by the disease on September 30. Within a week, the cholera began to spread through the surrounding neighborhood, and eventually through the entire nation. By the time the epidemic wound down, two years later, 50,000 people were dead.

Snow recognized immediately that this sequence of events posed a severe challenge to the opponents of the contagion model. The coincidence was simply too much for the miasma theory to bear. Two cases of cholera in a single room in the space of a week might be compatible with the miasma model, if one believed that the room itself contained some kind of noxious agent that poisoned its inhabitants. But it was stretching matters beyond belief to suggest that the room should suddenly become prone to those poisonous vapors the very day it was occupied by a sailor traveling from a city besieged by the disease. As Snow would later write: "Who can doubt that the case of John Harnold, the seaman from Hamburgh, mentioned above, was the true cause of the malady in Blenkinsopp, who came, and lodged, and slept, in the only room in all London in which there had been a case of true Asiatic cholera for a number of years? And if cholera be communicated in some instances, is there not the strongest probability that it is so in the others—in short, that similar effects depend on similar causes?"

But Snow also recognized the weakness of the contagionist argument. The same doctor attended both Harnold and Blenkinsopp, spending multiple hours in the room with them during the rice-water phase of the disease. And yet he remained free of the disease. Clearly, the cholera was not communicated through sheer proximity. In fact, the most puzzling element of the disease was that it seemed capable of traveling across city blocks, skipping entire houses in the process. The subsequent cases in Horsleydown erupted a few doors down from Harnold's original lodging house. You could be in the same room with a patient near death and emerge unscathed. But, somehow, you could avoid direct contact altogether with the infected person and yet still be seized with the cholera, simply because you lived in the same neighborhood. Snow grasped that solving the mystery of cholera would lie in reconciling these two seemingly contradictory facts.

We do not know if Snow hit upon the solution to this riddle sometime in the months that followed the initial 1848 outbreak, or if perhaps the solution had long lingered in the back of his mind, a hunch that had first taken shape more than a decade before, as he tended to the dying miners in Killingworth as a young surgeon's apprentice. We do know that in the weeks after the Horsleydown outbreak, as the cholera began its fatal march through the wider city and beyond, Snow embarked on a torrid stretch of inquiry: consulting with chemists who had studied the rice-water stools of cholera victims, mailing requests for information from the water and sewer authorities in Horsleydown, devouring accounts of the great epidemic of 1832. By the middle of 1849, he felt confident enough to go public with his theory. Cholera, Snow argued, was caused by some as-yet-unidentified agent that victims ingested, either through direct contact with the waste matter of other sufferers or, more likely, through drinking water that had been contaminated with that waste matter. Cholera was contagious, yes, but not in the way smallpox was contagious. Sanitary conditions were crucial to fighting the disease, but foul air had nothing to do with its transmission. Cholera wasn't something you inhaled. It was something you swallowed.

Snow built his argument for the waterborne theory around two primary studies, both of which showcased talents that would prove to be crucial five years later, during the Broad Street outbreak. In late July of 1849, an outbreak of cholera killed about twelve people living in slum conditions on Thomas Street in Horsleydown. Snow made an exhaustive inspection of the site and found ample evidence to support his developing theory. All twelve lived in a row of connected cottages called the Surrey building, which shared a single well in the courtyard they faced. A drainage channel for dirty water ran alongside the front of the houses, connecting to an open sewer at the end of the courtyard. Several large cracks in the drain allowed water to flow directly into the well, and during summer storms, the entire courtyard would flood with fetid water. And so a single case of cholera would quickly spread through the entire Surrey building population.

The layout of the Thomas Street flats provided Snow with an ingenious control study for his inquiry. The Surrey building backed onto a set of houses that faced another courtyard known as Truscott's Court. These abodes were every bit as squalid as the Surrey building, with the exact same demographic makeup of poor working families living within them. For all intents and purposes, they shared the same environment, save one crucial difference: they got their water from different sources. During the two-week period that saw the deaths of a dozen residents in the Surrey building, only one person perished in Truscott's Court, despite the fact that both groups lived within yards of each other. If the miasma were responsible for the outbreak, why would one squalid, impoverished group suffer ten times the loss of the one living next door?

The Thomas Street outbreak showcased Snow's on-the-ground investigative skills, his eye for the details of transmission patterns, sanitary habits, even architecture. But Snow also surveyed the outbreak from the bird's-eye view of citywide statistics. During his research, Snow had amassed an archive of information about the various companies that supplied water to the city, and that study had revealed a striking fact: that Londoners living south of the Thames were far more likely to drink water that had originated in the river as it passed through Central London. Londoners living north of the river drank from a variety of sources: some companies piped in water from the Thames above Hammersmith, far from the urban core; some drew from the New River in Hertfordshire to the north; others from the River Lea. But the South London Water Works drew its supply from the very stretch of the river where most of the city's sewers emptied. Anything that was multiplying in the city's intestinal tracts would be more likely to find its way into the drinking water of South London. If Snow's theory of cholera was on the mark, Londoners living below the Thames should have been significantly more prone to the disease than those living above.

Snow next surveyed the tables of cholera death that had been compiled by William Farr, London's registrar-general. What he found there followed the pattern that the water-supply routes predicted: of the 7,466 deaths in the metropolitan area during the 1848–1849 epidemic, 4,001 were located south of the Thames. That meant that the per capita casualty rate was near eight per thousand—three times that of the central city. In the growing suburbs of West and North London, the death rate was just above one per thousand. For the miasmatists, who were inclined to blame those death rates on the foul air of the working-class neighborhoods south of the river, Snow could point to the neighborhoods of the East End, which were probably the most destitute and overcrowded of any in the city. And yet their death rate was exactly half that of the area south of the Thames.

Whether you looked at the evidence on the scale of an urban courtyard or on the scale of entire city neighborhoods, the same pattern repeated itself: the cholera seemed to segment itself around shared water supplies. If the miasma theory were right, why would it draw such arbitrary distinctions? Why would the cholera devastate one building but leave the one next door unscathed? Why would one slum suffer twice the losses as a slum with arguably worse sanitary conditions?

Snow introduced his theory of cholera in two forms during the second half of 1849: first as a self-published thirty-one-page monograph, *On the Mode and Communication of Cholera,* intended for his immediate peers in the medical community, and then as an article in the *London Medical Gazette,* targeted at a slightly wider audience. Shortly after the publication, a country doctor named William Budd published an essay that came to similar conclusions about cholera's waterborne transmission, though Budd left open the possibility that some cases of cholera might be transmitted through the atmosphere, and he claimed, erroneously, to have identified the cholera agent in the form of a fungus growing in contaminated water supplies. Budd would later make an observation regarding the waterborne transmission of typhoid, for which he is now best known. But Snow's cholera theory had beaten Budd's to the presses by a month, and it did not include the false lead of fungal agents or of atmospheric transmission.

The reaction to Snow's argument was positive but skeptical. "Dr. Snow deserves the thanks of the profession for endeavouring to solve the mystery of the communication of cholera," a reviewer wrote in the *London Medical Gazette*. But Snow's case studies had not convinced: "[They] furnish no proof whatever of the correctness of his views." He had convincingly demonstrated that the South London neighborhoods were more at risk for cholera than the rest of the city, but it did not necessarily follow that the water in those neighborhoods was responsible for the disparity. Perhaps there was special toxicity to the air in those zones of the city that was absent in the slums to the north. Perhaps cholera was contagious, and thus the cluster of cases in South London simply reflected the chain of infection thus far; if the initial cases had unfolded differently, perhaps the East End would have been attacked more grievously, and South London left relatively unscathed. There was a correlation between water supply and cholera—that much Snow had convincingly proved. But he had not yet established a cause.

The *Gazette* did suggest one scenario that might settle the matter convincingly:

> The *experimentum crucis* would be, that the water conveyed to a distant locality, where cholera had been hitherto unknown, produced the disease in all who used it, while those who did not use it, escaped.

That passing suggestion stayed with Snow for five long years. As his anesthesia practice expanded, and his prominence grew, he continued to follow the details of each cholera outbreak, looking for a scenario that might help prove his theory. He probed, and studied, and waited. When word arrived of a terrible outbreak in Golden Square, not ten blocks from his new offices on Sackville Street, he was ready.

So many casualties in such a short stretch of time suggested a central contaminated water source used by large numbers of people. He needed to get samples of the water while the epidemic was still at full force. And so he made the journey across Soho, into the belly of the beast.

Snow's expectation was that contaminated water would have a cloudiness to it that was visible to the naked eye. But his initial glance at the Broad Street water surprised him; it was almost entirely clear. He drew samples from the other pumps in the area: Warwick Street, Vigo Street, Brandle Lane, and Little Marlborough Street. All were murkier than the Broad Street water. The Little Marlborough Street sample was worst of all. As he drew the water there, a handful of local residents on the street remarked that the pump water was notoriously poor—so poor, in fact, that many of them had taken to walking the extra blocks to Broad Street for their drinking water.

As Snow hurried back to his home on Sackville Street, he turned over the clues in his mind. Perhaps the Broad Street pump was not the culprit after all, given the lack of particles in the water. Perhaps one of the other pumps was the culprit? Or perhaps some other force was at work here? He would have a long night ahead of him, analyzing the samples, taking notes. He knew an outbreak of this magnitude could supply the linchpin for his argument. It was just a matter of finding the right evidence, and figuring out how to present that evidence in a way that would persuade the skeptics. Snow may well have been the only soul in Soho that day who found in the outbreak a glimmer of hope.

Snow didn't realize it at the time, but as he walked home that Sunday night, the basic pattern of that *experimentum crucis* suggested five years before in the *London Medical Gazette* was finally taking shape, miles away from Broad Street, in the greenery of Hampstead.

Susannah Eley had fallen ill earlier in the week after drinking her regular supply of Broad Street water, dutifully shipped to her by her children in Soho. By Saturday she was dead, followed on Sunday by her niece, who had returned to her home in Islington after visiting with her aunt. As Snow reviewed the pump-water samples in his microscope, Susannah Eley's servant, who had also consumed a glass of Broad Street water, remained locked in a life-or-death struggle with the disease.

Not one other case of cholera in Hampstead would be recorded for weeks.

IT'S ENTIRELY LIKELY THAT HENRY WHITEHEAD PASSED JOHN Snow on the streets of Soho that early evening. The young curate had toiled through another exhausting day, and was still making rounds well after the sun had set. Whitehead had begun the day with a feeling of hope; the fact that the streets seemed less chaotic made him wonder whether the outbreak was abating. A few of his initial visits offered reason for hope as well: the Waterstone girl had improved, and her father, having lost an otherwise perfectly healthy wife and daughter in less than two days' time, had begun consoling himself with the thought that life indeed might be worth living if his one remaining daughter somehow survived. Whitehead shared his upbeat assessment with a few of his colleagues on the sidewalk, and found some agreement.

But the quiet proved misleading: the streets were more tranquil because so much suffering was going on behind the shutters. In the end, another fifty would die over the course of the day. And new cases continued to appear at an alarming clip. When Whitehead returned to the Waterstones' at the end of the day, he found the sister continuing her improvement. But in the room next door, the girl's father was in the throes of cholera's initial attack. Life might well be worth living if his daughter survived, but the decision might not be up to him after all.

When Whitehead finally returned to his quarters at the end of that punishing day, he poured a glass of brandy and water, and found himself thinking of the Waterstones' ground-floor quarters. He had encountered the gossip that had been circulating in the past day, folk wisdom that would eventually find its way into the papers in the coming weeks: the residents of upper floors were dying at a more dramatic rate than those living on ground or parlor floors. There was a socioeconomic edge to this contention, one that reverses the traditional upstairs/downstairs division of labor: in Soho at the time, the bottom floors were more likely to be occupied by owners, with the upper floors rented out to the working poor. An increased death rate in the upper floors would suggest a fatal vulnerability in the constitution or sanitary habits of the poor. The notion, in its crude and haphazard way, was a version of Snow's tale of two buildings in Horsleydown: put two groups of people in close proximity, and if one group turns out to be significantly more vulnerable than the other, then some additional variable must be at work. For Snow, of course, the variable was water supply. But for the upstairs/downstairs rumor mill, the difference was class. A better sort of people lived on the ground floors—no wonder they were more likely to fight off the disease.

But as Whitehead reviewed his experiences over the preceding days, those easy assumptions began to wither in his mind. Yes, it did seem as though more people were dying on upper floors, but far more people *lived* on upper floors. And the Waterstones were clear evidence that the disease could assault ground-floor dwellers with impunity. Whitehead didn't have numbers in front of him, but he had a hunch from his anecdotal experiences that the lower floors had in fact been deadlier per capita over the previous forty-eight hours. It was certainly a fact worth investigating—if the pestilence ever moved on from Golden Square long enough to investigate anything.

Fifteen blocks away, on Sackville Street, John Snow was contemplating statistics as well. He had already sketched out a plan to ask William Farr for an early look at the mortality numbers. Perhaps there would be something in the distribution of deaths that would point to a contaminated water supply. Like Whitehead, Snow recognized that his work among the suffering of Golden Square had only begun. Whatever numbers William Farr provided him would have to be supplemented with local investigation. The longer he waited, the more difficult that investigation would become, if only because so many of the witnesses were dying.

Snow and Whitehead shared one other common experience that night. They both spent those last ruminating hours in the company of water drawn from the Broad Street pump. Snow was analyzing it in his home laboratory, his vision dimmed by the low light of candles. The young curate, however, had used the water in a different way, more recreational than empirical: he had mixed the water with a thimble of brandy and swallowed it.

WILLIAM FARR

Monday, September 4

THAT IS TO SAY, JO HAS NOT YET DIED

THE BRIGHT LATE-SUMMER SUN THAT ROSE OVER LONDON that Monday revealed a ghost town in the streets around Golden Square. Most who hadn't fallen ill, or who weren't tending to the fallen, had fled. Many of the storefronts remained closed for the day. A terrible gloom hovered over the Eley Brothers factory: more than two dozen laborers had been seized with the cholera, and news had arrived of Susannah Eley's death. (Little did the Eley brothers realize that their devotion to their mother had been instrumental in her demise.) The wife of Mr. G—the tailor who had been among the first to succumb—had herself collapsed the night before.

A few odd islands appeared in this sea of devastation. At the Lion Brewery, a hundred feet down Broad Street from the pump, work continued with a strange semblance of normalcy. Not one of the eighty laborers there had perished yet. The cholera continued to spare the tenements of Green's Court, despite their filthy, overcrowded quarters. Among the five hundred destitute residents at the St. James Workhouse on Poland Street, only a handful had come down with the disease, while the comparatively well-to-do houses that surrounded it had lost half their inhabitants in the space of three days.

But every time the Reverend Whitehead thought he saw reason for hope, another tragedy would arrive to dampen his natural optimism. When he returned to the Waterstones on Monday, he found that the lively, intelligent daughter he had long admired—whose health had taken a turn for the better the day before—had suffered a sudden relapse and died during the night. The few remaining family members were attempting to conceal the death from the girl's father, who continued his own struggle with the disease.

Whitehead began to hear talk spreading among his parishioners blaming the outbreak on the new sewers that had been constructed in recent years. The residents were whispering that the excavations had disturbed the corpses buried there during the Great Plague of 1665, releasing infectious miasma into the neighborhood's air. It was a kind of haunting, couched in the language of pseudoscience: the dead of one era's epidemic returning, centuries later, to destroy the settlers who had dared erect homes above their graves. The irony was that the terrified residents of Golden Square had it half right: those new sewers were in fact partly responsible for the outbreak that was devastating the city. But not because the sewers had disturbed a three-hundred-year-old graveyard. The sewers were killing people because of what they did to the water, not the air.

Other distortions and half-truths circulated between the neighborhood and the wider city. The folklore spread in part because the communication system of London in the middle of the nineteenth century was a strange mix of speed and sluggishness. The postal service was famously efficient, closer to e-mail than the appropriately nicknamed "snail mail" of today; a letter posted at nine a.m. would reliably find its way to its recipient across town by noon. (The papers of the day were filled with aggrieved letters to the editor complaining about a mailing that took all of six hours to find its destination.) But if person-to-person communication was shockingly swift, mass communication was less reliable. Newspapers were the only source of daily information about the wider state of the city, but for some reason the Broad Street outbreak went unmentioned for nearly four days in the city's main papers. One of the very first reports appeared in the weekly paper the *Observer,* though it greatly underestimated the magnitude of the attack: "It is said that Friday night will long be remembered by the inhabitants of Silver-street and Berwick-street. Seven persons were in good health on Friday night, and on Saturday morning they were all dead. Throughout the night people were running here and there for medical aid. It seemed as if the whole neighbourhood was completely poisoned."

With the newspapers largely silent on the topic, word of the terrible plague in Soho trickled out through the amplifying networks of gossip. Rumors began circulating that the entire neighborhood had been wiped out, that some new strain of cholera was killing people within minutes, that the dead were lying uncollected in the streets. More than a few Golden Square residents who worked outside the area left because their employers demanded they abandon their homes immediately.

The information channels were unreliable in both directions. In the belly of the beast, the terrified citizens of Soho traded stories: that the epidemic had struck Greater London with equal ferocity; that hundreds of thousands were dying; that the hospitals were overloaded beyond imagination.

But not all the locals had succumbed to abject fear. As he made his rounds, Whitehead found himself musing on an old saying that invariably surfaced during plague times: "Whilst pestilence slays its thousands, fear slays its tens of thousands." But if cowardice somehow made one more vulnerable to the ravages of the disease, Whitehead had seen no evidence of it. "The brave and the timid [were] indiscriminately dying and indiscriminately surviving," he would later write. For every terrified soul who fell victim to the cholera, there was another equally frightened survivor.

Fear might not have been a contributing factor in the spread of disease, but it had long been a defining emotion of urban life. Cities often began as an attempt to ward off outside threats—fortified by walls, protected by guards—but as they grew in size, they developed their own, internal dangers: disease, crime, fire, along with the "soft" dangers of moral decline, as many believed. Death was omnipresent, particularly for the working class. One study of mortality rates from 1842 had found that the average "gentleman" died at forty-five, while the average tradesman died in his mid-twenties. The laboring classes fared even worse: in Bethnal Green, the average life expectancy for the working poor was sixteen years. These numbers are so shockingly low because life was especially deadly for young children. The 1842 study found that 62 percent of all recorded deaths were of children under five. And yet despite this alarming mortality rate, the population was expanding at an extraordinary clip. Both the burial grounds *and* the streets were filling up with children. That contradictory reality explains, in part, the centrality of children in the Victorian novel, particularly in Dickens. There was, for the Victorians, something singularly charged about the idea of innocent children being exposed to the diseased squalor of the city, a notion that is, interestingly, almost entirely absent from French novels of the same period. When Dickens introduces the vagabond child Jo in *Bleak House,* his language implicitly references the dismal child-mortality statistics of the day: "Jo lives—that is to say, Jo has not yet died—in a ruinous place known to the like of him by the name of Tom-all-Alone's. It is a black, dilapidated street, avoided by all decent people, where the crazy houses were seized upon, when their decay was far advanced, by some bold vagrants who after establishing their own possession took to letting them out in lodgings." The phrasing captures the dark reality of urban poverty: to live in such a world was to live with the shadow of death hovering over your shoulder at every moment. To live was to be not dead yet.

From our vantage point, more than a century later, it is hard to tell how heavily that fear weighed upon the minds of individual Victorians. As a matter of practical reality, the threat of sudden devastation—your entire extended family wiped out in a matter of days—was far more immediate than the terror threats of today. At the height of a nineteenth-century cholera outbreak, a thousand Londoners would often die of the disease in a matter of weeks—out of a population that was a quarter the size of modern New York. Imagine the terror and panic if a biological attack killed four thousand otherwise healthy New Yorkers over a twenty-day period. Living amid cholera in 1854 was like living in a world where urban tragedies on that scale happened week after week, year after year. A world where it was not at all out of the ordinary for an entire family to die in the space of forty-eight hours, children suffering alone in the arsenic-lit dark next to the corpses of their parents.

Outbreaks had an ominous preamble, too. Newspapers would track the disease's progress through the harbors and trading towns of Europe, as it marched relentlessly across the Continent. When cholera first appeared in New York City in the summer of 1832, it attacked the city from the north: arriving first in Montreal via ships originating in France, the disease spent a month snaking along the trade routes of upstate New York toward the city, then floating straight down the Hudson. Every few days the papers would announce that the cholera had taken another step; when it eventually arrived, in early July, almost half the city had escaped to the countryside, creating traffic jams that resembled the Long Island Expressway on a modern-day Fourth of July weekend. The *New York Evening Post* reported:

> The roads, in all directions, were lined with well-filled stage coaches, livery coaches, private vehicles and equestrians, all panic struck, fleeing from the city, as we may suppose the inhabitants of Pompeii or Reggio fled from those devoted places, when the red lava showered down upon their houses, or when the walls were shaken asunder by an earthquake.

The popular fear of cholera was amplified by the miasma theory of its transmission. The disease was both invisible and everywhere: seeping out of gulley holes, looming in the yellowed fog along the Thames. The courage of those who stayed to fight the disease—or investigate its origins—is all the more impressive in this light, since simply breathing in the vicinity of an outbreak was assumed by almost everyone to be risking death. John Snow had at least the courage of his convictions to rely on: if the cholera was in the water, then venturing into the Golden Square neighborhood at the height of the epidemic posed no grave threat, as long as he refrained from drinking the pump water during his visits. The Reverend Whitehead had no such theory to allay his fears as he spent hour after hour sitting in the presence of the sick, and yet not once in his writing about the Broad Street outbreak is there mention of his own private dread.

It is hard to peer behind that absence, to extract the real truth of Whitehead's mental state: Was he terrified but still compelled into action by his faith and his sense of duty to the parish? And compelled, by pride, to avoid mention of his terror in his subsequent writing? Or did his religious convictions help him ward off his fear, as Snow's scientific convictions helped him? Or had he simply acclimated to the constant presence of death?

Certainly some process of acclimation must have been at work. Otherwise, it is hard to imagine how Londoners survived such dangerous times without being paralyzed by terror. (Not all escaped the anxiety, however; witness the prevalence of hysterics in so much Victorian fiction. The corset may not have been the only culprit behind all those fainting spells.) The spike in cases of posttraumatic stress disorder experienced by big-city dwellers after 9/11 is conventionally attributed to a sudden rise in danger thanks to terrorist threat, particularly in iconic urban centers like New York, London, and Washington, D.C. But the long view suggests that this account has it exactly backward. We feel fear more strongly because our safety expectations have risen so dramatically over the past hundred years. Even with its higher crime rate, New York City in its debauched nadir of the 1970s was a vastly safer place to live than Victorian London. During the epidemics of the late 1840s and the 1850s, a thousand Londoners would typically die of cholera in a matter of weeks—in a city a quarter the size of present-day New York—and the deaths would barely warrant a headline. And so, as shocking as those numbers seem to us now, they may not have provoked the same mortal panic that they trigger today. The literature—both public and private—of the nineteenth century is filled with many dark emotions: misery, humiliation, drudgery, rage. But terror does not quite play the role that one might expect, given the body count.

Far more prevalent was another feeling: that things could not continue at this pace for long. The city was headed toward some kind of climactic breaking point that would likely undo the tremendous growth of the preceding century. This was a profoundly dialectical feeling, a thesis giving rise to an antithesis, the city's success eventually breeding the very conditions of its destruction, like the "avenging ghost" in Dickens' eulogy for the opium-addicted scrivener in *Bleak House*.

London, of course, had a long history of offending social critics, as in this cheery description from Scottish physician George Cheyne, written at the end of the eighteenth century:

> The infinite number of Fires, Sulphurous and Bituminous, the vast expense of Tallow and foetid Oil in Candles and Lamps, under and above the Ground, the clouds of Stinking Breathes and Perspirations, not to mention the ordure of so many diseas'd, both intelligent and unintelligent animals, the crouded Churches, Church Yards and Bury Places, with the putrifying Bodies, the Sinks, Butchers Houses, Stables, Dunghills, etc. and the necessary Stagnation, Fermentation, and mixture of Variety of all Kinds of Atoms, and more than sufficient to putrefy, poison and infect the Air for Twenty Miles around it, and which in Time must alter, weaken, and destroy the healthiest of Constitutions.

Part of this disgust can be attributed to the fact that the classical distinction between the metropolis and the industrial towns to the north—one the center of commerce and services, the others of industry and manufacturing—was not nearly as clearly defined as it eventually became in the late 1800s. At the end of the eighteenth century, London had more steam engines than all of Lancashire, and it remained the manufacturing center of England until 1850. Factories like the Eley brothers' would be dramatically out of place next to the shops and residences of today's London, but they were an ordinary sight (not to mention smell) in 1854.

Accounts of London's repulsive conditions inevitably imagined the city as a unified organism, a sprawling, cancerous body laid out along the Thames. In prose that sounds more like a medical diagnosis than an economic forecast, Sir Richard Phillips predicted in 1813 that

> the houses will become too numerous for the inhabitants, and certain districts will be occupied by beggary and vice, or become depopulated. This disease will spread like an atrophy in the human body, and ruin will follow ruin, till the entire city is disgusting to the remnant of the inhabitants; at length the whole becomes a heap of ruins: Such have been the causes of decay of all overgrown cities. Nineveh, Babylon, Antioch, and Thebes are become heaps of ruins. Rome, Delphi, and Alexandria are partaking the same inevitable fate; and London must some time from similar causes succumb under the destiny of every thing human.

It is here that the modern urban mind confronts what may be the largest gap separating it from the Victorian worldview. In a very practical sense, no one had ever tried to pack nearly three million people inside a thirty-mile circumference before. The metropolitan city, as a concept, was still unproven. It seemed entirely likely to many reasonable citizens of Victorian England—as well as to countless visitors from overseas—that a hundred years from now the whole project of maintaining cities of this scale would have proved a passing fancy. The monster would eat itself.

Most of us don't harbor doubts of this scale today, at least where cities are concerned. We worry about other matters: the epic shantytowns of Third World megacities; the terror threats; the environmental impact of a planet industrializing at such a dramatic rate. But most of us accept without debate the long-term viability of human settlements with populations in the millions, or tens of millions. We know it can be done. We just haven't figured out how to ensure that it is done well.

And so, in projecting back to the mind-set of a Londoner in 1854, we have to remember this crucial reality: that a sort of existential doubt lingered over the city, a suspicion not that London was flawed, but that the very idea of building cities on the scale of London was a mistake, one that was soon to be corrected.

IF LONDON WAS SUCH A RANK, OVERCROWDED SEWER IN THE first half of the nineteenth century, then why did so many people decide to move there? No doubt there were those who savored the energy and stimulus of the city, its architecture and parks, its coffeehouse sociability, its intellectual circles. (Wordsworth's *Prelude* even included a paean to shopping: "the string of dazzling wares, / Shop after shop, with symbols, blazoned names, / And all the tradesman's honours overhead.") But for every intellectual or aristocrat moving to the city for its cosmopolitan flavor, there were a hundred mudlarks and costermongers and night-soil men who must have had a very different aesthetic response to the city.

The tremendous growth of London—like the parallel explosions of Manchester and Leeds—was a riddle that could not be explained by simply adding up decisions made by large numbers of individual humans. This was, ultimately, what perplexed and horrified so many onlookers at the time: the sense that the city had taken on a life of its own. It was the product of human choice, to be sure, but some new form of collective human choice where the collective decisions were at odds with the needs and desires of its individual members. If you had somehow polled the population of Victorian England and asked them if stacking two million people inside a thirty-mile circumference was a good idea, the answer would have been a resounding no. But somehow, the two million showed up anyway.

That perplexity gave rise to an intuitive sense that the city itself was best understood as a creature with its own distinct form of volition, greater than the sum of its parts: a monster, a diseased body—or, most presciently, Wordsworth's "anthill on the plain." (The unplanned but complex engineering of ant colonies display a number of striking similarities to human cities.) The observers of the time were detecting a phenomenon that we now largely take for granted: that "mass" behavior can often diverge strikingly from the desires of the individuals that make up the mass. Even if you had the time to write it all down, you couldn't tell the story of a city as an endless series of individual biographies. You had to think of collective behavior as something distinct from individual choice. To capture the city in its entirety, you had to move one level up the chain, to the bird's-eye view. Henry Mayhew famously took to a hot-air balloon in any attempt to take in the entirety of the city from a single vantage point, but found, to his dismay, that the "monster city . . . stretched not only to the horizon on either side, but far away into the distance."

The sense, then, of London as a monstrous, cancerous presence focused not merely on the smell or the overcrowding; it also included the uncanny feeling that, somehow, humans themselves were not in control of the urbanization process. In this the Victorians were grasping at an underlying reality that they were only partially able to understand. Cities tend to be imagined in terms of their streets, or markets, or buildings (or, to the twentieth-century mind, their skylines). But they are ultimately shaped by flows of energy. The hunter-gatherers or the early agriculturists couldn't have formed a city of the size and density of 1850s London (much less today's São Paulo) even if they had wanted to. To sustain a population of a million people—to keep them fed alone, much less power their SUVs or subways or refrigerators—you need a massive supply of stored energy to keep all those bodies alive. Small bands of hunter-gatherers collected enough energy, if they were lucky, to sustain small bands of hunter-gatherers. But when the Fertile Crescent's proto-farmers began planting fields of cereal grains, they dramatically increased the energy available to their settlements, allowing populations to swell into the thousands, and, in the process, creating density levels that had never been seen before among the primates, much less the humans. Soon, positive feedback loops emerged: more people working the fields increased the food supply, which allowed more people to work the fields, and so on. Eventually, these first agricultural societies achieved what may still be the sine qua non of civilization: a large class of people liberated from the day-to-day problem of finding a new source of food. Cities were suddenly populated by a class of *consumers,* free to worry about other pressing matters: new technologies, new modes of commerce, politics, professional sports, celebrity gossip.

That same process drove the explosion of metropolitan London after 1750. Three related developments had triggered an unprecedented intensification of the energy flowing through the capital. First, the "improvements" of agrarian capitalism, where the dotted, irregular system of the feudal English countryside gave way to rationalistic agriculture; second, the energy unleashed by the coal and steam power of the Industrial Revolution; third, the dramatic increase in the portability of that energy thanks to the railway system. For millennia, most cities had been bound inexorably to the natural ecosystem that lay outside their walls: the energy flowing through the fields and forests around them established a population ceiling they couldn't grow beyond. London in 1854 had shot through those ceilings, because the land itself was being farmed more efficiently, because new forms of energy had been discovered, and because shipping and railway networks had greatly expanded the distance that energy could travel. The Londoner enjoying a cup of tea with sugar in 1854 was drawing upon a vast global energy network with each sip: the human labor of the sugarcane plantations in the West Indies and the newly formed tea plantations in India; the solar energy in those tropical realms that allowed those plants to flourish; the oceanic energy of the trade currents, and the steam power of the railway engine; the fossil fuels powering the looms in Lancashire, making fabrics that helped fund the entire trade system.

The great city, then, could not be understood as an artifact of human choice. It was much closer to a natural, organic process—less like a building that has been deliberately constructed and more like a garden erupting into full bloom with the arrival of spring—a mix of human planning and the natural developmental patterns that emerge with increasing energy supplies. Several decades ago, the physicist Arthur Iberall proposed that patterns of human organization could be understood as the social equivalent of the patterns formed by molecules in response to changing energy states. A collection of water molecules follows a reliable pattern of transformations depending on how much energy is injected into the system: in low-energy situations, it takes on the crystal form of ice, while high-energy infusions transform liquid water into a gas. The dramatic shifts from one state to another are called phase transitions, or bifurcations. Iberall observed that human societies appeared to cycle through comparable phase transitions, as the energy harnessed by the society increased: moving from the gaseous state of roaming hunter-gathers, to the more settled configuration of agrarian farming, to the crystalline density of the walled city. When the supply of surplus energy spiked, thanks to the slave labor and transportation networks of the Roman Empire, the city of Rome itself surged to more than a million people, and dozens of towns connected to that network reached populations in the hundreds of thousands. But when the imperial system crumbled, the energy supply dried up, and the cities of Europe vaporized in a matter of centuries. By the year 1000—right around the time the next great energy revolution was stirring—Rome had been whittled down to a mere 35,000 people, one-thirtieth of its former glory.

Growing a city of three million from less than a million a century before required more than just increased energy inputs, however. It also required an immense population base that was willing to move from the country to the city. As it happened, the enclosure movement that dominated so much of British rural life during the 1700s and early 1800s created a huge surge in mobility by disrupting the open-field farming system that had been in place since medieval times. Hundreds of thousands, if not millions, of tenant farmers who had resided in rural hamlets, living off common land, suddenly found their ancient lifestyle upended by a long wave of privatization. Those newly free-floating laborers became another, equally essential, energy source for the Industrial Revolution, filling its cities and coketowns with a nearly inexhaustible supply of cheap labor. In a sense, the Industrial Revolution would have never happened if two distinct forms of energy had not been separated from the earth: coal and commoners.

The dramatic increase of people available to populate the new urban spaces of the Industrial Age may have had one other cause: tea. The population growth during the first half of the eighteenth century neatly coincided with the mass adoption of tea as the de facto national beverage. (Imports grew from six tons at the beginning of the century to eleven thousand at the end.) A luxury good at the start of the century, tea had become a staple even of working-class diets by the 1850s. One mechanic who provided an account of his weekly budget to the *Penny Newsman* spent almost fifteen percent of his earnings on tea and sugar. He may have been indulging in it for the taste and the salutary cognitive effects of caffeine, but it was also a healthy lifestyle choice, given the alternatives. Brewed tea possesses several crucial antibacterial properties that help ward off waterborne diseases: the tannic acid released in the steeping process kills off those bacteria that haven't already perished during the boiling of the water. The explosion of tea drinking in the late 1700s was, from the bacteria's point of view, a microbial holocaust. Physicians observed a dramatic drop in dysentery and child mortality during the period. (The antiseptic agents in tea could be passed on to infants through breast milk.) Largely freed from waterborne disease agents, the tea-drinking population began to swell in number, ultimately supplying a larger labor pool to the emerging factory towns, and to the great sprawling monster of London itself.

Do not mistake these multiple trends—the energy flows of metropolitan growth, the new taste for tea, the nascent, half-formed awareness of mass behavior—for mere historical background. The clash of microbe and man that played out on Broad Street for ten days in 1854 was itself partly a consequence of each of these trends, though the chains of cause and effect played out on different scales of experience, both temporal and spatial. You can tell the story of the Broad Street outbreak on the scale of a few hundred human lives, people drinking water from a pump, getting sick and dying over a few weeks, but in telling the story that way, you limit its perspective, limit its ability to convey a fair account of what really happened, and, more important—*why* it happened. Once you get to why, the story has to widen and tighten at the same time: to the long *durée* of urban development, or the microscopic tight focus of bacterial life cycles. These are causes, too.

There is a lovely symmetry that comes from telling the story this way, because a city and a bacterium are each situated at the very extreme boundaries of the shapes that life takes on earth. Viewed from space, the only recurring evidence of man's presence on this planet are the cities we build. And in the night view of the planet, cities are the only thing going at all, geologic *or* biologic. (Think of those pulsing clusters of streetlights, arranged in the chaotic, but still recognizable patterns of *real* human settlement patterns, and not the clean, imperial geometry of political borders.) With the exception of the earth's atmosphere, the city is life's largest footprint. And microbes are its smallest. As you zoom in past the scale of the bacterium and the virus, you travel from the regime of biology to the regime of chemistry: from organisms with a pattern of growth and development, life and death, to mere molecules. It is a great testimony to the connectedness of life on earth that the fates of the largest and the tiniest life should be so closely dependent on each other. In a city like Victorian London, unchallenged by military threats and bursting with new forms of capital and energy, microbes were the primary force reigning in the city's otherwise runaway growth, precisely because London had offered *Vibrio cholerae* (not to mention countless other species of bacterium) precisely what it had offered stockbrokers and coffeehouse proprietors and sewer-hunters: a whole new way of making a living.

So the macro-growth of the urban superorganism and the microscopic subtleties of the bacterium are both essential to the events of September 1854. In some cases, the chains of cause and effect are obvious ones. Without the population densities and the global connectivity of industrialization, cholera might not have been as devastating in England, and thus might not have attracted Snow's investigative skills in the first place. But in other places, the causal chains are more subtle, though no less important to the story. The bird's-eye view of the city, the sense of the urban universe as a system, as a mass phenomenon—this *imaginative* breakthrough is as crucial to the eventual outcome of the Broad Street epidemic as any other factor. To solve the riddle of cholera you had to zoom out, look for broader patterns in the disease's itinerary through the city. When health matters are at stake, we now call this wide view epidemiology, and we have entire university departments devoted to it. But for the Victorians, the perspective was an elusive one; it was a way of thinking about patterns of social behavior that they had trouble intuitively grasping. The London Epidemiological Society had been formed only four years before, with Snow as a founding member. The basic technique of population statistics—measuring the incidence of a given phenomenon (disease, crime, poverty) as a percentage of overall population size—had entered the mainstream of scientific and medical thought only in the previous two decades. Epidemiology as a science was still in its infancy, and many of its basic principles had yet to be established.

At the same time, the scientific method rarely intersected with the development and testing of new treatments and medicines. When you read through that endless stream of quack cholera cures published in the daily papers, what strikes you most is not that they are all, almost without exception, based on anecdotal evidence. What's striking is that they never apologize for this shortcoming. They never pause to say, "Of course, this is all based on anecdotal evidence, but hear me out." There's no *shame* in these letters, no awareness of the imperfection of the method, precisely because it seemed eminently reasonable that local observation of a handful of cases might serve up the cure for cholera, if you looked hard enough.

But cholera couldn't be studied in isolation. It was as much a product of the urban explosion as the newspapers and coffeehouses where it was so uselessly anatomized. To understand the beast, you needed to think on the scale of the city, from the bird's-eye view. You needed to look at the problem from the perspective of Henry Mayhew's balloon. And you needed a way to persuade others to join you there.

THAT WIDER PERSPECTIVE IS WHAT JOHN SNOW FOUND himself searching for by noon on Monday. He had reexamined his samples from the Soho wells in the light of day and found nothing suspicious in the Broad Street water. As he delivered chloroform to a patient of a nearby dentist who was performing a tooth extraction, he pondered the outbreak still raging a few blocks away. The more he thought about it, the more convinced he became that the water supply must have been contaminated somehow. But how to prove it? The water alone might not be sufficient, since he didn't even know what he was looking for. He had a theory about cholera's routes of transmission and its effects on the body. But he had no idea what the agent that caused cholera was exactly, much less how to identify it.

Ironically, just a few days before Snow had unsuccessfully attempted to see any telltale signs of cholera in the water, an Italian scientist at the University of Florence had discovered a small, comma-shaped organism in the intestinal mucosa of a cholera victim. It was the first recorded sighting of *Vibrio cholerae,* and Filippo Pacini published a paper that year describing his findings, under the title "Microscopical Observations and Pathological Deductions on Cholera." But it was the wrong time for such a discovery: the germ theory of disease had not yet entered mainstream scientific thought, and cholera itself was largely assumed by the miasmatists to be some kind of atmospheric pollution, not a living creature. Pacini's paper was ignored, and *V. cholerae* retreated back into the invisible kingdom of microbes for another thirty years. John Snow would go to his grave never learning that the cholera agent he had spent so many years pursuing had been identified during his lifetime.

The fact that Snow had no idea what cholera looked like under the microscope didn't stop him from doing further tests on the water. After his appointment with the dentist, he returned to draw more samples from the Broad Street pump. This time he saw small white particles in the water. Back in his lab, he ran a quick chemistry experiment, which reported an unusually high presence of chlorides. Encouraged, he took the sample to a colleague, Dr. Arthur Hassall, whose skill with the microscope Snow had long admired. Hassall reported that the particles had no "organized structure," which led him to believe they were the remnants of decomposed organic matter. He also saw a host of oval-shaped life-forms—Hassall called them "animalculae"—presumably feeding on the organic substances.

So the Broad Street water was not as pure as he had originally thought. But still, there was nothing in Hassall's analysis that pointed definitively to the presence of cholera. If he was going to crack this case, the solution wouldn't be found under the microscope, on the scale of particles and animalculae. He needed to approach the problem from the bird's-eye view, on the scale of neighborhoods. He would try to find the killer through an indirect route: by looking at patterns of lives and deaths on the streets of Golden Square.

As it turned out, Snow had already spent much of the past year thinking about cholera from this perspective. After his first publications at end of the 1840s had failed to persuade the medical authorities of his waterborne theory, Snow had continued looking for evidence supporting this theory. He followed outbreaks in Exeter, Hull, and York from afar. He read William Farr's *Weekly Returns of Birth and Deaths* the way the rest of the population devoured the installments of *Bleak House* and *Hard Times.* Each outbreak of the disease offered a new configuration of variables, a new pattern—and thus the possibility for a new kind of experiment, one that would unfold in the streets and cemeteries rather than in Snow's crowded flat. In this, Snow developed a strangely symbiotic relationship with *V. cholerae*: he needed the disease to flourish to have a shot at conquering it. The quiet years between 1850 and 1853, during which the cholera was largely dormant in England, were good years for the health of the nation. But they were unproductive ones for Snow the investigator. When the cholera returned with a vengeance in 1853, he threw himself into Farr's *Weekly Returns* with extra zeal, scanning the charts and tables for clues.

In Farr, Snow had the closest thing to an ally in the existing medical establishment. In many ways their lives had followed parallel paths. Born to poor Shropshire laborers five years before Snow, Farr had trained as a doctor in the 1830s but went on to revolutionize the use of statistics in public health in the following decade. He had joined the newly created Registrar-General's Office in 1838, a few months after his first wife had died of that other nineteenth-century killer, tuberculosis. Farr had been hired to track the most elemental of demographic trends: the number of births, deaths, and marriages in England and Wales. Over time, though, he had refined the statistics to track more subtle patterns in the population. "Bills of Mortality" dated back to the plague years of the 1600s, when clerks first began recording the names and parishes of the dead. But Farr recognized that these surveys could be far more valuable to science if they included additional variables. He waged a long campaign to persuade physicians and surgeons to report a cause of death wherever possible, drawing upon a list of twenty-seven fatal diseases. By the mid-1840s, his reports tallied deaths not only by disease, but also by parish, age, and occupation. For the first time, doctors and scientists and health authorities had a reliable vantage point from which to survey the broad patterns of disease in British society. Without Farr's *Weekly Returns,* Snow would have been stuck in the street-level view of anecdote, hearsay, and direct observation. He might still have been able to build a theory of cholera on his own, but it would have been almost impossible to persuade anyone else of its validity.

Farr was a man of science, and shared Snow's belief in the power of statistics to shed light on medical riddles. But he also shared many assumptions with the miasma camp, and he used the number-crunching of the *Weekly Returns* to reinforce those beliefs. Farr thought that the single most reliable predictor of environmental contamination was elevation: the population living in the putrid fog that hung along the riverbanks were more likely to be seized by the cholera than those living in the rarefied air of, say, Hampstead. And so, after the 1849 outbreak, Farr began tabulating cholera deaths by elevation, and indeed the numbers seemed to show that higher ground was safer ground. This would prove to be a classic case of correlation being mistaken for causation: the communities at the higher elevations tended to be less densely settled than the crowded streets around the Thames, and their distance from the river made them less likely to drink its contaminated water. Higher elevations were safer, but not because they were free of miasma. They were safer because they tended to have cleaner water.

Farr was not entirely opposed to Snow's theory. He seems to have entertained the idea that the cholera was somehow originating in the murky waters of the Thames, and then rising into the smoggy air above the river as some kind of poisonous vapor. He had clearly followed Snow's publications and presentations closely over the years, and engaged the theory on occasion in the editorials that would sometimes accompany the *Weekly Returns.* But he remained unconvinced by the purely waterborne theory. He also suspected that Snow would have a difficult time proving his theory. "To measure the effects of good or bad water supply," Farr editorialized in November of 1853, "it is requisite to find two classes of inhabitants living at the same level, moving in equal space, enjoying an equal share of the means of subsistence, engaged in the same pursuits, but differing in this respect,—that one drinks water from Battersea, the other from Kew. . . . But of such *experimenta crucis* the circumstances of London do not admit."

Snow must have taken that last line as a slap in the face, having heard the exact same Latinate phrase used against him after the publication of his original cholera monograph four years before. Yet despite his skepticism, Farr had been intrigued enough by Snow's waterborne theory to add a new category to his *Weekly Returns.* In addition to tracking the age and sex and elevation of the cholera victims, Farr would now track one additional variable: where they got their water.

THE SEARCH FOR UNPOLLUTED DRINKING WATER IS AS OLD as civilization itself. As soon as there were mass human settlements, waterborne diseases like dysentery became a crucial population bottleneck. For much of human history, the solution to this chronic public-health issue was not purifying the water supply. The solution was to drink alcohol. In a community lacking pure-water supplies, the closest thing to "pure" fluid was alcohol. Whatever health risks were posed by beer (and later wine) in the early days of agrarian settlements were more than offset by alcohol's antibacterial properties. Dying of cirrhosis of the liver in your forties was better than dying of dysentery in your twenties. Many genetically minded historians believe that the confluence of urban living and the discovery of alcohol created a massive selection pressure on the genes of all humans who abandoned the hunter-gatherer lifestyle. Alcohol, after all, is a deadly poison and notoriously addictive. To digest large quantities of it, you need to be able to boost production of enzymes called alcohol dehydrogenases, a trait regulated by a set of genes on chromosome four in human DNA. Many early agrarians lacked that trait, and thus were genetically incapable of "holding their liquor." Consequently, many of them died childless at an early age, either from alcohol abuse or from waterborne diseases. Over generations, the gene pool of the first farmers became increasingly dominated by individuals who could drink beer on a regular basis. Most of the world's population today is made up of descendants of those early beer drinkers, and we have largely inherited their genetic tolerance for alcohol. (The same is true of lactose tolerance, which went from a rare genetic trait to the mainstream among the descendants of the herders, thanks to the domestication of livestock.) The descendants of hunter-gatherers—like many Native Americans or Australian Aborigines—were never forced through this genetic bottleneck, and so today they show disproportionate rates of alcoholism. The chronic drinking problem in Native American populations has been blamed on everything from the weak "Indian constitution" to the humiliating abuses of the U.S. reservation system. But their alcohol intolerance mostly likely has another explanation: their ancestors didn't live in towns.

Ironically, the antibacterial properties of beer—and all fermented spirits—originate in the labor of other microbes, thanks to the ancient metabolic strategy of fermentation. Fermenting organisms, like the unicellular yeast fungus used in brewing beer, survive by converting sugars and carbohydrates into ATP, the energy currency of all life. But the process is not entirely clean. In breaking down the molecules, the yeast cells discharge two waste products—carbon dioxide and ethanol. One provides the fizz, the other the buzz. And so in battling the health crisis posed by faulty waste-recycling in human settlements, the proto-farmers unknowingly stumbled across the strategy of consuming the microscopic waste products generated by the fermenters. They drank the waste discharged by yeasts so that they could drink their own waste without dying in mass numbers. They weren't aware of it, of course, but in effect they had domesticated one microbial life-form in order to counter the threat posed by other microbes. The strategy persisted for millennia, as the world's civilizations discovered beer, then wine, then spirits—until tea and coffee arrived to offer comparable protection against disease without employing the services of fermenting microbes.

But by the middle of the nineteenth century, in England at least, water was finding a role for itself in the urban diet. Starting in the mid-1700s, a growing patchwork of privately owned water pipes began snaking their way through the city, supplying the wealthiest Londoners with running water in their homes (or, in some cases, depositing the water in a cistern near their house). It is difficult to overestimate the revolutionary impact of this advance. So many of the household conveniences of modern life—the dishwashers and washing machines and toilets and showers—depend on a reliable supply of water. Just being able to pour yourself a glass from a faucet in your home would have been miraculous to the Londoners who first experienced it.

By the mid-1800s, the loose assortment of small firms running the water pipes had consolidated into roughly ten major firms, each with its own protected turf in the city. The New River Water Company supplied the city proper, while the Chelsea Water Company piped to the West End. South of the Thames, two companies controlled the area: Southwark and Vauxhall (otherwise known as S&V), and Lambeth. Many of these companies—including S&V and Lambeth—had intake pipes within the tidal reach of the Thames. The water they supplied their customers was therefore contaminated by the raw waste of the city, thanks to the growing network of sewers that emptied into the increasingly foul river. Even the most ardent miasmatist could find something offensive in that arrangement, and so in the early 1850s, Parliament passed legislation ordering that all London's water companies had to move their intake pipes above the tidewater mark by August 1855. S&V chose to delay its move to the very last minute, continuing to draw from Battersea, but Lambeth switched its waterworks to the far cleaner supply at Thames Ditton in 1852.

Snow had been following the water companies since his early investigation of 1849 and had already been tracking the results of

Lambeth's move. But the real breakthrough came in the form of a footnote in the November 26 edition of the *Weekly Returns.* Below the cholera deaths for South London, Farr had appended this seemingly innocuous line: "In three cases . . . the same districts are supplied by two companies."

That minor bit of infrastructure trivia would have immediately struck Snow as a tremendous opportunity. A population all living in the same space, at the same elevation, divided between two water supplies, one rank with the sewage of the city, the other comparatively pure. Farr's footnote had inadvertently supplied Snow with his *experimenta crucis.*

All Snow needed was a further breakdown: a record of deaths originating in houses that had been supplied with S&V water, and deaths in houses supplied by Lambeth. If Snow's theory was right, there should be a disproportionate fatality rate in the S&V homes, despite the fact that they existed side by side with the Lambeth homes. Their elevation and air quality would all be the same—only the water would be different. Even economic status and upbringing would be taken out of the equation, since the rich and poor were just as likely to choose one water supply over the other. It would be the Thomas Street flats all over again: shared environment, different water. But this time the scale would be immense: thousands of lives, not dozens. As Snow would eventually describe it:

> The experiment . . . was on the grandest scale. No fewer than three hundred thousand people of both sexes, of every age and occupation, and of every rank and station, from gentlefolks down to the very poor, were divided into two groups without their choice, and, in most cases, without their knowledge; one group being supplied with water containing the sewage of London, and, amongst it,

> whatever might have come from the cholera patients, the other group having water quite free from such impurity.

But the *experimentum crucis* would prove to be thornier than Snow anticipated. Farr's original report had looked only at the level of entire districts, but Snow now divided the original data into subdistricts organized by water supplier. Twelve of them relied on water from S&V, while three drank Lambeth water exclusively. And indeed, the disparity between the two groups in terms of cholera deaths was pronounced: roughly 1 in 100 died in the S&V subdistricts, while not a single person had died of cholera among the 14,632 Lambeth drinkers. An unbiased observer might have been persuaded by those numbers, but Snow realized his audience required more, primarily because the subdistricts served by Lambeth alone were relatively well-to-do suburbs, in contrast to the smog-bound industrial zones that S&V serviced. Once the miasmatists had a look at the different neighborhoods, Snow knew his case would dissolve in a heartbeat.

And so the experiment would rise and fall on the sixteen remaining subdistricts that received *both* S&V and Lambeth water. If Snow could find a breakdown of cholera deaths within those districts along the lines of water supplier, he might well have conclusive proof of his theory, enough perhaps to turn the tide against the miasma model. But those numbers turned out to be elusive ones, because the pipes in those sixteen subdistricts were so promiscuously interlinked that it was impossible to tell from a given address which water company serviced it. If Snow wanted to disentangle the water supply of the sixteen, he would have to rely on old-fashioned shoe leather to do it. He would have to knock on every door mentioned in Farr's account, and inquire where people procured their water.

It is worth pausing for a second to reflect on Snow's willingness to pursue his investigation this far. Here we have a man who had reached the very pinnacle of Victorian medical practice—attending on the queen of England with a procedure that he himself had pioneered—who was nonetheless willing to spend every spare moment away from his practice knocking on hundreds of doors in some of London's most dangerous neighborhoods, seeking out specifically those houses that had been attacked by the most dread disease of the age. But without that tenacity, without that fearlessness, without that readiness to leave behind the safety of professional success and royal patronage, and venture into the streets, his "grand experiment"—as Snow came to call it—would have gone nowhere. The miasma theory would have remained unchallenged.

Yet descending to the street-level scale of direct interviews ultimately proved unsatisfactory as well. Many residents had no idea where their water came from. Either the bills were paid by a distant landlord, or they had paid no notice to the company name when they last received an invoice and weren't in the habit of keeping old paperwork around. The visible pipes were so jumbled that even direct inspection couldn't reveal whether it was Lambeth or S&V water running into each house.

And so Snow's inquiry had to venture down to an even smaller scale to track its quarry. The grand experiment that had begun with the bird's-eye view of hundreds of thousands of lives would ultimately revolve around molecules invisible to the unaided human eye. In the course of his investigation, Snow had noticed that S&V water consistently contained about four times as much salt as Lambeth water. A simple test in his home lab could determine which company had supplied the water. From that point on, anytime Snow encountered a resident who had no idea who provided the water they were drinking, Snow would simply draw a small vial of water, mark it with the address, and analyze the contents when he returned home.

SO THIS IS WHERE JOHN SNOW FOUND HIMSELF PROFESSIONALLY when the cholera arrived at Golden Square: splitting his days between chloroform and shoe leather, leading a double life of celebrated anesthesiologist and South London investigator. By late August of 1854, the essential components of his grand experiment were in place, and the early returns were promising. All he needed was a few more weeks pounding the pavement of Kennington, Brixton, and Waterloo, and perhaps a few more weeks beyond that to tally up the numbers. When the cholera first struck a few blocks from his flat, the temptation to ignore the outbreak and continue with his grand experiment must have been tremendous. He had been chasing this thread for almost a year now, ever since Farr's footnote had caught his eye. Another outbreak would be a distraction. But as word spread of the severity of the attack, Snow recognized that the Golden Square case might prove as revealing as his South London inquiry. By the end of Monday—with his water tests inconclusive, and the epidemic still raging around him—he was knocking on doors again, this time in his own neighborhood. All around him, the signs of devastation were inescapable. The *Observer* would later report: "In Broad-street, on Monday evening, when the hearses came round to remove the dead, the coffins were so numerous that they were put on top of the hearses as well as the inside. Such a spectacle has not been witnessed in London since the time of the plague."

EDWIN CHADWICK

Tuesday, September 5

ALL SMELL IS DISEASE

THE FIRST SOLID CAUSE FOR HOPE BEGAN TO FILTER through the neighborhood Tuesday morning. For the first time in four days, Henry Whitehead let himself believe that this terrible visitation might finally be passing. The wife of Mr. G, the tailor, had died that morning, but for every new death, Whitehead could point to another dramatic recovery. The servant woman he had been tending to since Friday had risen from what she had assumed would be her deathbed, her pallor much improved. Two adolescents—a boy and girl—had also turned the corner, much to the delight of their remaining family. All three of them attributed their recovery to one thing: they had consumed large quantities of water from the Broad Street pump since falling ill. The speed and intensity of their recovery made an impression on Whitehead that would linger in his mind through the coming weeks.

In the late-morning hours, a small, formal parade of government officials, the members of the General Board of Health, arrived in Golden Square to tour the scene of the outbreak. The most notable thing about the procession was its leader: the Board's new president, Sir Benjamin Hall, who had replaced the pioneering but controversial Edwin Chadwick a month earlier, prompting the *Morning Chronicle* to observe dryly that the incoming president was coming to the job "with one great advantage to his favor—his predecessors managed to accumulate upon themselves so much unpopularity that he has little to fear from invidious contrasts."

As the officials walked through Dufours Place and Broad Street, small bands of surviving locals appeared on the sidewalk to express their gratitude for the Board's appearance, their spirits also cheered by the sense that the outbreak was subsiding. The Board's secretary released an account of the visit to the major papers, most of which obligingly reprinted it, including in their copy the self-congratulatory line: "The Guardians are acting most energetically, and every credit is due to them." But it was harder to specify what those actions were exactly, however energetic they might have been. The outbreak might have been diminishing, but it was still taking lives at a monstrous clip. More than five hundred residents of the Golden Square neighborhood had died in five days, and another seventy-six had fallen ill the day before. The *Times* itself was circumspect in describing what the Board was actually doing to battle the outbreak, beyond mentioning plans to form a committee to investigate it. The Board would eventually have a role to play in the Broad Street drama, but for the moment its actions were mostly theater.

The one intervention the Board of Health had made would have been immediately and viscerally evident to anyone walking through the neighborhood: the streets had been soaked with chloride of lime, and the smell of bleach was omnipresent, blocking out the usual stench of urban waste. In this one intervention, Edwin Chadwick's influence on the Board lingered past his tenure as its head. The lime had been deployed to battle Chadwick's lifelong nemesis, the sanitary curse he had built a career fulminating against, and the one he would go to his grave believing in: miasma.

IT IS NEARLY IMPOSSIBLE TO OVERSTATE THE IMPACT THAT Edwin Chadwick's life had on the modern conception of government's proper role. From 1832, when he was first appointed to the Poor Law Commission, through his landmark 1842 study of sanitation among the laboring classes, through his tenure as commissioner of the sewers in the late 1840s, to his final run at the helm of the General Board of Health, Chadwick helped solidify, if not outright invent, an ensemble of categories that we now take for granted: that the state should directly engage in protecting the health and well-being of its citizens, particularly the poorest among them; that a centralized bureaucracy of experts can solve societal problems that free markets either exacerbate or ignore; that public-health issues often require massive state investment in infrastructure or prevention. For better or worse, Chadwick's career can be seen as the very point of origin for the whole concept of "big government" as we know it today.

Most of us today accept that the broad movements of Chadwick's campaigns were ultimately positive ones. You have to be a committed libertarian or anarchist to think that the government shouldn't be building sewers or funding the Centers for Disease Control or monitoring the public water supply. But if Chadwick's long-term legacy was a progressive one, his short-term track record, as of 1854, was more complicated. No doubt he had done more than anyone alive to focus attention on the shameful condition of the industrial poor, and to mobilize forces to correct those problems. But some of the most significant programs he put in place ended up having catastrophic effects. Thousands upon thousands of cholera deaths in the 1850s can be directly attributed to decisions that Chadwick made in the decade before. This is the great irony of Chadwick's life: in the process of inventing the whole idea of a social safety net, he unwittingly sent thousands of Londoners to an early grave.

How could such noble aspirations lead to such devastating results? In Chadwick's case, there is a simple explanation: he insisted, to the point of obstinacy, on following his nose. The air of London was killing Londoners, he claimed, and thus the route to public health had to begin with removing noxious smells. He expressed this notion most famously—and most comically—in his 1846 testimony to a parliamentary committee investigating the problem of London's sewage: "All smell is, if it be intense, immediate acute disease; and eventually we may say that, by depressing the system and rendering it susceptible to the action of other causes, all smell is disease."

WITH FEW EXCEPTIONS, THE PROBLEMS THAT THE EARLY Victorians wrestled with are still relevant more than a century later. These are the standard social questions that you'll encounter in any textbook account of the period: How can a society industrialize in a humane way? How can a government rein in the excesses of the free market? To what extent should working people be allowed to negotiate collectively?

But there was another debate that ran alongside those more austere themes, one that has not received as much attention in the seminar rooms or the biographies. It's true enough that the Victorians were grappling with heady issues like utilitarianism and class consciousness. But the finest minds of the era were also devoted to an equally pressing question: *What are we going to do with all of this shit?*

The extent of London's excrement problem was universally agreed upon. Chadwick's influential 1842 study had laboriously recounted the repellent state of waste disposal in the city. Letter writers to the *Times* and other papers harped on the topic endlessly. A survey in 1849 examined 15,000 homes, and found that almost 3,000 had offensive smells from bad drainage, while a thousand had "privities [*sic*] and water-closets in a very offensive state." One in twenty had human waste piling up in the cellar.

Many prominent reformers saw economic waste in all that fecal matter. Using human excrement as fertilizer in the greenlands around city centers was an ancient practice, but it had never been attempted with the waste of two million people. Hyperfertile soils would inevitably result if such a project were carried out, the evangelists claimed. One expert projected a fourfold increase in food production. A proposal in 1843 argued for the construction of cast-iron sewers that would transport waste all the way to Kent and Essex.

Few were as rhapsodic on the subject as Henry Mayhew, who saw in waste recycling an escape route from the Malthusian limits on population growth: "If what we excrete plants secrete—if what we exhale they inspire—if our refuse is their food—then it follows that to increase the population is to increase the quantity of manure, while to increase the manure is to augment the food of plants, and consequently the plants themselves. If the plants nourish us, we at least nourish them."

As was typical of Mayhew, this circle-of-life philosophizing quickly gave way to a frenzy of numerical calculation:

> According to the average of the returns, from 1841 to 1846, we are paying two millions every year for guano, bone-dust, and other foreign fertilizers of our soil. In 1845, we employed no fewer than 683 ships to bring home 220,000 tons of animal manure from Ichaboe alone; and yet we are every day emptying into the Thames 115,000 tons of a substance which has been proved to be possessed of even greater fertilizing powers. With 200 tons of the sewage that we are wont to regard as refuse, applied to the irrigation of one acre of meadow land, seven crops, we are told, have been produced in the year, each of them worth from 6l. [6 pounds sterling] to 7l.; so that, considering the produce to have been doubled by these means, we have an increase of upwards of 20l. per acre per annum effected by the application of that refuse to the surface of our fields. This return is at the rate of 10l. for every 100 tons of sewage; and, since the total amount of refuse discharged into the Thames from the sewers of the metropolis is, in round numbers, 40,000,000 tons per annum, it follows that, according to such estimate, we are positively wasting 4,000,000l. of money every year.

This sort of bookkeeping remained an essential subgenre in the political debate for decades to come. One scholar testified before Parliament in 1864 that the value of London's sewage was "equal to the local taxation of England, Ireland, and Scotland." The Victorians were literally flushing money down the toilet—or, worse, leaving it to decay in the cellar.

Edwin Chadwick, too, was a great believer in the economic bounty that lay trapped in London's sewage. A document he helped produce in 1851 argued that fertilizing the countryside with London's waste would cause land values to quadruple. He also entertained an aquatic version of the theory, arguing that delivering fresh feces in an expedient manner to England's waterways would produce larger fish.

But for Chadwick and other social reformers of the period, the primary reason to deal with London's rising tide of excrement had to do with health, not economics. Not everyone went as far as Chadwick's conviction that all smell was disease, but most agreed that the vast quantities of waste decomposing in the cellars and the streets of the city were literally poisoning the air. If merely taking a stroll down the sidewalk could overwhelm you with the putrid stench of human waste, something clearly had to be done.

The solution was straightforward enough, at least in theory. London needed a citywide sewage system that could remove waste products from houses in a reliable and sanitary fashion. It would require a massive engineering effort, but a country that had built a national rail network in a matter of decades and spearheaded the Industrial Revolution could handle a project on that scale. The problem was one of jurisdiction, not execution. The urban infrastructure of early Victorian London was governed by a byzantine assortment of local boards that had been assembled over the centuries by more than two hundred separate acts of Parliament. Paving or lighting the streets, building drains and sewers—these were all acts overseen by local commissioners with almost no citywide coordination. One three-quarter-mile stretch of the Strand was overseen by nine separate paving boards. To take on a project as epic as building an integrated metropolitan sewer system would require more than engineering genius and backbreaking labor. It would need a revolution in the power dynamics of city life. The bottom-up, improvised recycling of the scavengers would have to give way to the master planner.

In this, Edwin Chadwick was perfectly cast for the role. Brusque and strong-willed to the point of rudeness, Chadwick was in many ways a Victorian rendition of Robert Moses (that is, if Moses had lost his grip on New York City's power structure halfway through his career and spent the last thirty years of his life editorializing from the sidelines). A devout Utilitarian and friend of Jeremy Bentham, Chadwick had spent the thirties helping to create—and then, partially, clean up—the national mess that was the Poor Law Acts of 1832–1834. But by the 1840s he had grown increasingly obsessed with sanitation issues, and his crusades ultimately culminated in the passing of the Public Health Act of 1848, which established the three-member General Board of Health, Chadwick at its helm. But the bill with the most dramatic short-term impact on London's health would be the Nuisances Removal and Contagious Diseases Prevention Act, also passed in 1848, after years of Chadwick's campaigning. "Nuisance" in this case meant, effectively, one thing: human waste. For a few years, new buildings had been required to drain into the existing sewer system, but the "cholera bill"—as it was conventionally referred to—was the first to require sewer connections from *existing* structures. For the first time, the law had something to say about people opting to fill their old cellars with "great heaps of turds," as Samuel Pepys put it in a 1660 journal entry. Though of course the law didn't quite express it that way—choosing a more delicate, if prolix, language to describe the problem:

> [Any] Dwelling House or Building in any City, Town, Borough, Parish, or Place within or over which the Jurisdiction or Authority of the Town Council, Trustees, Commissioners, Guardians, Officers of Health, or other Body to whom such Notice is given, extends, is in such a filthy and unwholesome Condition as to be a Nuisance and injurious to the Health of any Person, or that upon any Premises within such Jurisdiction of Authority there is any foul or offensive Ditch, Gutter, Drain, Privy, Cesspool, or Ashpit, or any Ditch, Gutter, Drain, Privy, Cesspool, or Ashpit kept or constructed

> so as to be a Nuisance to or injurious to the Health of any Person, or that upon any such Premises Swine, or any Accumulation of Dung, Manure, Offal, Filth, Refuse, or other Matter or Thing, are or is kept so as to be a Nuisance to or injurious to the Health of any Person, or that upon such Premises . . .

To abide by these new laws, though, you needed somewhere to put all that "manure, offal, and filth." You needed working sewers. London actually had an ancient drainage system that had evolved around a dozen creeks and small rivers that continue to flow beneath the city to this day. (The largest waterway, the Fleet River, runs beneath Farringdon Road, emptying into the Thames under Blackfriars Bridge.) Parliamentary bills governing the construction of new sewers date back to the days of Henry VIII. Historically, however, London's sewers had been designed to carry off the city's surface water. Until 1815, it was illegal to discharge raw waste into the sewers. If your cesspool was overflowing, you called the night-soil men. This system resulted in some foul-smelling cellars, but it left the waters of the Thames remarkably pristine, with a bustling fisherman's trade working the river between Greenwich and Putney Bridge. But as the city's population exploded, and as more and more houses discharged their waste into the existing sewers, the quality of the Thames water declined at an alarming rate. What's more, the sewers themselves began to clog, leading to the occasional underground explosion of methane gas.

Chadwick's work in the 1840s and early fifties had the perverse effect of exacerbating this problem, both through his position as head of the Board of Health and his seat on the newly formed Metropolitan Commission of Sewers. There was much squabbling and drafting of plans for expanding the city's sewage system, but nothing practical was done for years, until a brilliant engineer named Joseph Bazalgette took charge of the project. In the meantime, the primary focus was on eliminating cesspools. As Bazalgette would later report: "Within a period of about six years, thirty thousand cesspools were abolished, and all house and street refuse was turned into the river." Several times a year, the Commission's engineers would offer enthusiastic reports documenting just how much waste had been extracted from the city's houses and deposited in the river: 29,000 cubic yards in the spring of 1848, growing quickly to 80,000 cubic yards the following winter. In the space of about thirty-five years, the Thames had been transformed from a fishing ground teeming with salmon to one of the most polluted waterways in the world—all in the name of public health. As the builder Thomas Cubbitt observed wryly: "The Thames is now made a great cesspool instead of each person having one of his own."

Herein lies the dominant irony of the state of British public health in the late 1840s. Just as Snow was concocting his theory of cholera as a waterborne agent that had to be ingested to do harm, Chadwick was building an elaborate scheme that would deliver the cholera bacteria directly to the mouths of Londoners. (A modern bioterrorist couldn't have come up with a more ingenious and far-reaching scheme.) Sure enough, the cholera returned with a vengeance in 1848–1849, the rising death toll neatly following the Sewer Commission's cheerful data on the growing supply of waste deposited in the river. By the end of the outbreak, nearly 15,000 Londoners would be dead. The first defining act of a modern, centralized public-health authority was to poison an entire urban population. (There is some precedent to Chadwick's folly, however. During the plague years of 1665–1666, popular lore had it that the disease was being spread by dogs and cats. The Lord Mayor promptly called for a mass extermination of the city's entire population of pets and strays, which was dutifully carried out by his minions. Of course, the plague turned out to be transmitted via the rats, whose numbers grew exponentially after the sudden, state-sponsored demise of their only predators.)

Why would the authorities go to such lengths to destroy the Thames? All the members of these various commissions were fully aware that the waste being flushed into the river was having disastrous effects on the quality of the water. And they were equally aware that a significant percentage of the population was drinking that water. Even without a waterborne theory of cholera's origin, it seems like madness to celebrate the ever-increasing tonnage of human excrement being flushed into the water supply. And, indeed, it was a kind of madness, the madness that comes from being under the spell of a Theory. If all smell was disease, if London's health crisis was entirely attributable to contaminated air, then any effort to rid the houses and streets of miasmatic vapors was worth the cost, even if it meant turning the Thames into a river of sewage.

CHADWICK MAY HAVE BEEN THE MOST INFLUENTIAL MIASmatist of his age, but he had plenty of illustrious company. The other great social crusaders of the age were equally convinced of the connection between foul air and disease. In 1849, the *Morning Chronicle* sent Henry Mayhew to the heart of the cholera epidemic, in the Bermondsey neighborhood south of the river. The account eventually published deserves its own distinct journalistic genre—olfactory reporting:

> On entering the precincts of the pest island, the air has literally the smell of a graveyard, and a feeling of nausea and heaviness comes over any one unaccustomed to imbibe the musty atmosphere. It is not only the nose, but the stomach, that tells how heavily the air is

> loaded with sulphuretted hydrogen; and as soon as you cross one of the crazy and rotting bridges over the reeking ditch, you know, as surely as if you had chemically tested it, by the black colour of what was once the white-lead paint upon the door-posts and window-sills, that the air is thickly charged with this deadly gas. The heavy bubbles which now and then rise up in the water show you whence at least a portion of the mephitic compound comes, while the open doorless privies that hang over the water side on one of the banks, and the dark streaks of filth down the walls where the drains from each house discharge themselves into the ditch on the opposite side, tell you how the pollution of the ditch is supplied.

The scientific establishment was equally anchored in the miasma theory. In September 1849, the *Times* ran a series of articles that surveyed the existing theories about cholera: "How is the cholera generated?—how spread? what is its modus operandi on the human frame? These questions are in every mouth," the paper observed, before taking a decidedly pessimistic stance on the question of whether they would ever be answered:

> These problems are, and will probably ever remain, among the inscrutable secrets of nature. They belong to a class of questions radically inaccessible to the human intelligence. What the forces are which generate phenomena we cannot tell. We know as little of the vital force itself as of the poison-forces which have the power to disturb or suppress it.

Despite this bleak forecast, the *Times* went on to survey the prevailing theories: a "telluric theory that supposes the poison to be an emanation from the earth"; an "electric theory" based on atmopheric conditions; the ozonic theory that attributed outbreaks to a deficiency of ozone in the air; a theory that blamed cholera on "putrescent yeast, emanations of sewers, graveyards, etc." The paper also mentioned a theory that maintained that the disease was spread by microscopic animalcules or fungi, though it downplayed its viability, claiming that the theory "failed to include all the observed phenomena."

The diversity of views is striking here—ozone, sewer emanations, electricity—but just as striking is the underlying commonality: all but one of the theories assume that the cholera is somehow being transmitted through the atmosphere. (Snow's waterborne theory, already a matter of public record, goes completely unmentioned.) The air was the key to the riddle of cholera, and indeed to most known diseases. Nowhere is the philosophy more pronounced than in the writings of the Victorian age's most beloved and influential medical figure, Florence Nightingale. Consider this passage from the beginning of her groundbreaking 1857 work *Notes on Nursing*:

> The very first canon of nursing, the first and the last thing upon which a nurse's attention must be fixed, the first essential to a patient, without which all the rest you can do for him is as nothing, with which I had almost said you may leave all the rest alone, is this: TO KEEP THE AIR HE BREATHES AS PURE AS THE EXTERNAL AIR, WITHOUT CHILLING HIM. Yet what is so little attended to? Even where it is thought of at all, the most extraordinary misconceptions reign about it. Even in admitting air into the patient's room or ward, few people ever think, where that air comes from. It may come from a corridor into which other wards are ventilated, from a hall, always unaired, always full of the fumes of gas, dinner, of various kinds of mustiness; from an underground kitchen, sink, washhouse, water-closet, or even, as I myself have had

> sorrowful experience, from open sewers loaded with filth; and with this the patient's room or ward is aired, as it is called—poisoned, it should rather be said.

With Nightingale, the problem is one of emphasis; there's obviously nothing wrong with ensuring that hospital rooms have fresh air in them. The problem arises when supplying clean air becomes the single most important task for the doctor or nurse, when the air is assumed to be a "poison" that has caused the patient's illness in the first place. Nightingale believed that cholera, smallpox, measles, and scarlet fever were all miasmatic in nature, and she recommended that schools, homes, and hospitals use a certain "air test," devised by the chemist Angus Smith, that detected organic materials in the air:

> If the tell-tale air test were to exhibit in the morning, both to nurses and patients, and to the superior officer going round, what the atmosphere has been during the night, I question if any greater security could be afforded against a recurrence of the misdemeanor.
>
> And oh, the crowded national school! where so many children's epidemics have their origin, what a tale its air-test would tell! We should have parents saying, and saying rightly, "I will not send my child to that school, the air-test stands at 'Horrid.'" And the dormitories of our great boarding schools! Scarlet fever would be no more ascribed to contagion, but to its right cause, the air-test standing at "Foul."
>
> We should hear no longer of "Mysterious Dispensations," and of "Plague and Pestilence," being "in God's hands," when, so far as we know, He has put them into our own. The little air-test would both betray the cause of these "mysterious pestilences," and call upon us to remedy it.

So often what is lacking in many of these explanations and prescriptions is some measure of humility, some sense that the theory being put forward is still unproven. It's not just that the authorities of the day were wrong about miasma; it's the tenacious, unquestioning way they went about being wrong. An investigator looking for holes in the theory could find them everywhere, even in the writings of the miasmatists themselves. The canary in the miasma coal mine should have been the sewer-hunters, who spent their waking hours exposed to the most noxious—sometimes even explosive—air imaginable. And yet, bizarrely, the canary seemed to be doing just fine, and Mayhew admits as much in one slightly puzzled passage in *London Labour and the London Poor*:

> It might be supposed that the sewer-hunters (passing much of their time in the midst of the noisome vapours generated by the sewers, the odour of which, escaping upwards from the gratings in the streets, is dreaded and shunned by all as something pestilential) would exhibit in their pallid faces the unmistakable evidence of their unhealthy employment. But this is far from the fact. Strange to say, the sewer-hunters are strong, robust, and healthy men, generally florid in their complexion, while many of them know illness only by name. Some of the elder men, who head the gangs when exploring the sewers, are between 60 and 80 years of age, and have followed the employment during their whole lives.

As Snow observed many times in his writings during the period, there were countless cases of groups sharing the exact same living environment, breathing the exact same air, who seemed to have entirely opposing responses to the allegedly poisonous vapors. If the miasma was truly killing off Londoners, it seemed to choose its victims in an entirely arbitrary fashion. And despite the fact that Chadwick and his commissions had made immense progress in eliminating the city's population of cesspools, the cholera had nonetheless come roaring back to devastate the city in 1853.

All of which begs the central question: Why was the miasma theory so persuasive? Why did so many brilliant minds cling to it, despite the mounting evidence that suggested it was false? This kind of question leads one to a kind of mirror-image version of intellectual history: not the history of breakthroughs and eureka moments, but instead the history of canards and false leads, the history of being wrong. Whenever smart people cling to an outlandishly incorrect idea despite substantial evidence to the contrary, something interesting is at work. In the case of miasma, that something involves a convergence of multiple forces, all coming together to prop up a theory that should have died out decades before. Some of those forces were ideological in nature, matters of social prejudice and convention. Some revolved around conceptual limitations, failures of imagination and analysis. Some involve the basic wiring of the human brain itself. Each on its own might not have been strong enough to persuade an entire public-health system to empty raw sewage into the Thames. But together they created a kind of perfect storm of error.

MIASMA CERTAINLY HAD THE FORCE OF TRADITION ON ITS side. The word itself is a derivation from the Greek term for pollution; the notion of disease being transmitted by poisoned air dates back to Greek medicine of the third century B.C. Hippocrates was so obsessed with air-quality issues that his medical tracts sometimes sound like instructions for a novice meteorologist. His treatise *On Air,*

Water, and Places begins: "Whoever wishes to investigate medicine properly, should proceed thus: in the first place to consider the seasons of the year, and what effects each of them produces for they are not at all alike, but differ much from themselves in regard to their changes. Then the winds, the hot and the cold, especially such as are common to all countries, and then such as are peculiar to each locality." (Farr would echo this philosophy centuries later: his *Weekly Returns* would invariably include a brief weather report, before getting to the body count.) Just about every epidemic disease on record has been, at one point or another, attributed to poisoned miasma. The word "malaria" itself derives from the Italian *mal aria,* or "bad air."

Miasma theories were eminently compatible with religious tradition as well. As one might expect from a man of the cloth, Henry Whitehead believed that the Golden Square outbreak was God's will, but he supplemented his theological explanation with a miasmatic one; he believed that "the atmosphere, all over the world, is at this time favourable to the production of a most formidable plague." To reconcile this hideous reality with the idea of a beneficent Creator, Whitehead had settled on what might later have been termed an ingeniously Darwinian explanation: that plagues were God's way of adapting the human body to global changes in the atmosphere, killing off thousands or millions, but in the process creating generations that could thrive in the new environment.

But tradition alone can't account for the predominance of the miasma theory. The Victorians who clung to it were in almost every other respect true revolutionaries, living in revolutionary times: Chadwick was inventing a whole new model for shaping public health; Farr transforming the use of statistics; Nightingale challenging countless received ideas about the role of women in professional life, as well as the practice of nursing. Dickens, Engels, Mayhew—these were not people naturally inclined to accept the status quo. In fact, they were all, in their separate ways, spoiling for a fight. So it's not sufficient to blame their adherence to the miasma theory purely on its long pedigree.

The perseverance of miasma theory into the nineteenth century was as much a matter of instinct as it was intellectual tradition. Again and again in the literature of miasma, the argument is inextricably linked to the author's visceral disgust at the smells of the city. The sense of smell is often described as the most primitive of the senses, provoking powerful feelings of lust or repulsion, triggering *mémoires involontaires.* (Proust's original madeleine-inspired reverie was triggered largely by taste, but the power of smell alone is a recurring theme of *In Search of Lost Time,* and of course smell is an essential component of taste.) Modern brain-imaging technology has revealed the intimate physiological connection between the olfactory system and the brain's emotional centers. In fact, the seat of many of those emotional centers—the limbic system—was once called the "rhinencephalon," literally "nose-brain" or "smell-brain." A 2003 study found that strong smells triggered activity in both the amygdala and the ventral insula. The amygdala is an evolutionarily ancient part of the brain, much older than the mammalian higher functions of the neocortex; raw instinctual responses to threats and emotionally charged stimuli emanate from the amygdala. The ventral insula appears to play an important role in biological urges, like hunger, thirst, and nausea, as well as in certain phobias. Both regions can be thought of as alarm centers of the brain; in humans, they possess the capacity to override the neocortical systems where language-based reasoning occurs. The brain scans in the 2003 study found that sharply unpleasant smells triggered disproportionately strong responses in both the amygdala and the ventral insula.

In lay terms, the human brain appears to have evolved an alert system whereby a certain class of extreme smells triggers an involuntary disgust response that effectively short-circuits one's ability to think clearly—and produces a powerful desire to avoid objects associated with the smell. It is easy to imagine the evolutionary pressures that would bring this trait into being. Once again microbes are at the center of the story. Eating meat or vegetation that has already begun the decomposition process poses a significant health risk, as does eating foods that have been contaminated with fecal matter—precisely because of the microbial life-forms that are doing the decomposing. Putrefying foods release several organic compounds into the air; they have names like putrescine and cadaverine. Bacteria recycling energy stored in fecal matter releases hydrogen sulfide into the air. Disgust at the scent of any of these compounds is as close to a universal human trait as we know. You can think of it as a form of evolutionary pattern recognition: over millions of years of evolution, natural selection hit upon the insight that the presence of hydrogen sulfide molecules in the air was a reasonably good predictor that microbial life-forms that could be dangerous if swallowed were nearby. And so the brain evolved a system for setting off an alarm whenever those molecules were detected. Nausea itself was a survival mechanism: it was better to void the contents of your stomach than run the risk that the smell was coming from the antelope you'd just finished eating.

But those telltale molecules—hydrogen sulfide, cadaverine—were *clues* pointing to a threat. They were not the threat itself. If you press your nose up against a decomposing banana or antelope, you might well make yourself vomit, but you won't contract a disease from the experience, however repulsive. Breathing in pure methane gas or hydrogen sulfide could kill you, of course, but bacterial decomposition doesn't release anywhere near enough of these gases to saturate the environment. In other words, methane and putrescine and cadaverine are the smoke. Microbes are the fire.

Basing the alarm system around smell made perfect sense for the environmental conditions of the hunter-gatherer lifestyle. The smell of decay and fecal waste was relatively rare in a world where humans lived in small roving bands; there were no sewers or dustheaps on the savannahs of Africa, precisely because the hunter-gatherers had such low population densities and such mobile lifestyles. You could just leave your waste behind and move on to a new spot; odds were, the bacteria would have recycled it all by the time another human returned. The alarm system of disgust likely evolved both because the threat posed by eating decaying organic matter was a serious one, and because the smell that signaled the presence of decaying matter was *unusual.* If the smell had been ubiquitous—if, say, some common African flower had begun emitting hydrogen sulfide from its blooms—then the human brain might have evolved another way of anticipating the presence of decaying food.

The trouble is that survival strategies optimized for a hunter-gatherer lifestyle play out differently in a modern city of two million people. Civilization had produced many transformations in the experience of human life: farms, wheels, books, railroads. But civilized life had another distinguishing feature: it was a lot smellier. Densely packed populations of people without modern waste-management systems produced powerfully repellent odors. When Mayhew describes his repulsion at the smell of hydrogen sulfide on the streets of Bermondsey, you can see in the passage a clash between three distinct epochs somehow struggling to share the same space: an industrial-era city with an Elizabethan-era waste-removal system as perceived by a Pleistocene-era brain.

The miasmatists had plenty of science and statistics and anecdotal evidence to demonstrate that the smells of London weren't killing people. But their gut instincts—or, more like it, their amygdalas—kept telling them otherwise. All of John Snow's detailed, rigorous analysis of the water companies and the transmission routes of the Horsleydown outbreak couldn't compete with a single whiff of the air in Bermondsey. The miasmatists were unable to override the alarm system that had evolved so many aeons before. They mistook the smoke for the fire.

MIASMA'S HEGEMONY HAD ONE OTHER BIOLOGICAL BASIS. Our noses are far more adept than our eyes at perceiving the very small. It takes only a few molecules of cadaverine attaching to the olfactory receptors in your upper nasal passages for you to become aware of the smell of decay. But your eyes are useless at the scale of molecules. In many respects, human visual perception is unrivaled among earth's life-forms—the legacy of a nocturnal mammal who needed to forage and hunt in the dark. But molecules remain several orders of magnitude below the threshold of human visual perception. We can't see most ordinary cells that those molecules build, even whole populations of cells. A hundred million *V. cholerae* floating in a glass of water would be invisible to the naked eye. Microscopes had been in use for more than two centuries, and while a few isolated researchers had caught a glimpse of microbes in their labs, the existence of a bacterial microcosmos was still the stuff of fantasy and conjecture for the mid-Victorian mind. But the stench of decomposition was all too real. Smelling was believing.

The miasma theory drew on other sources for its power as well. It was as much a crisis of imagination as it was pure optics. To build a case for waterborne cholera, the mind had to travel across scales of human experience, from the impossibly small—the invisible kingdom of microbes—to the anatomy of the digestive tract, to the routine daily patterns of drinking wells or paying the water-company bills, all the way up to the grand cycles of life and death recorded in the *Weekly Returns.* If you looked at cholera on any one of those levels, it retreated back into the haze of mystery, where it could be readily rolled back to the miasma theory, given the pedigree and influence of miasma's supporters. Miasma was so much less complicated. You didn't need to build a consilient chain of argument to make the case for miasma. You just needed to point to the air and say: *Do you smell that?*

And of course there were more than a few instances where the statistical evidence did in fact seem to stack the odds in miasma's favor. Neighborhoods with unsanitary water supplies generally suffered from poor air quality as well; many of them lay at the lower elevations that Farr relentlessly documented in his *Weekly Returns.* For every sewer-hunter living happily into his sixties, there were a hundred false positives dying in the low elevations of Bermondsey.

Raw social prejudice also played a role. Like the other great scientific embarrassment of the period—phrenology—the miasma theory was regularly invoked to justify all sorts of groundless class and ethnic biases. The air was poisoned, to be sure, but the matter of who fell ill, and what disease they suffered from, was determined by the constitution of each individual breathing in the air. So went Thomas Sydenham's internal-constitution theory of the epidemic, an eccentric hybrid of weather forecasting and medieval humorology. Certain atmospheric conditions were likely to spawn epidemic disease, but the nature of the diseases that emerged depended partly on a kind of preexisting condition, a constitutional susceptibility to smallpox, or influenza, or cholera. The distinction was often defined as one between exciting and predisposing causes. The exciting cause was the atmospheric condition that encouraged a certain kind of disease: a specific weather pattern that might lead to yellow fever, or cholera. The predisposing cause lay in the bodies of the sufferers themselves. That constitutional failing was invariably linked to moral or social failing: poverty, alcohol abuse, unsanitary living. One alleged expert argued in 1850: "The probability of an outburst or increase during [calm, mild] weather, I believed to be heightened on holidays, Saturdays, Sundays, and any other occasions where opportunities were afforded the lower classes for dissipation and debauchery."

The idea of one's internal constitution shaping the manifestation of disease was not just useful for affirming social prejudices about the moral depravity of the lower classes. It also helped paper over a massive hole in the theory itself. If the miasma seemed unusually capricious in its choice of victims for poison allegedly circulating in the atmosphere—if it killed off two housemates but left the remaining two unscathed despite the fact that they were all breathing the same air—the miasmatists could simply point to the differences in constitution between the victims and the survivors to explain the disparity. Although the poisonous vapors were distributed equally through the environment, each inner constitution possessed its own distinct vulnerability.

Like much of the reasoning that lay behind the miasma theory, the idea of an inner constitution was not entirely wrong; immune systems do vary from person to person, and some people may indeed be resistant to epidemic diseases like cholera or smallpox or plague. The scaffolding that kept miasma propped up for so long was largely made up of comparable half-truths, correlations mistaken for causes. Methane and hydrogen sulfide *were* in fact poisons, after all; they just weren't concentrated enough in the city air to cause real damage. People *were* more likely to die of cholera at lower elevations, but not for the reasons Farr imagined. And the poor *did* have higher rates of contagion than the well-to-do, but not because they were morally debauched.

Yet miasma had just as much to offer the liberals as it did the conservatives. Chadwick and Nightingale and Dickens were hardly bigots where the working classes were concerned. Miasma, for them, was not a public sign of the underclasses' moral failing; it was a sign of the deplorable conditions in which the underclasses had been forced to live. It seemed only logical that subjecting such an immense number of people to such deplorable living environments would have a detrimental effect on their health, and of course, the liberal miasmatists were entirely right in those basic assumptions. Where they went wrong was in assuming that the primary culprit lay in the air.

And so, on August 29, when the *Morning Chronicle* welcomed Benjamin Hall to his new job as president of the Board of Health, the editors included more than a few cutting remarks at the expense of Edwin Chadwick; yet they embraced the theory of miasma with both arms and urged the new president to continue the work of enforcing the Nuisances Removal and Contagious Diseases Prevention Act. There may be no clearer example of miasma's dark irony: on the very day that the outbreak in Golden Square was beginning, one of London's most prestigious papers was urging the Board of Health to accelerate its work poisoning the water supply.

MIASMA TURNS OUT TO BE A CLASSIC CASE OF WHAT FREUD, in another context, called "overdetermination." It was theory that drew its persuasive power not from any single fact but rather from its location at the intersection of so many separate but compatible elements, like a network of isolated streams that suddenly converges to form a river. The weight of tradition, the evolutionary history of disgust, technological limitations in microscopy, social prejudice—all these factors colluded to make it almost impossible for the Victorians to see miasma for the red herring that it was, however much they prided themselves on their Gradgrindian rationality. Every research paradigm, valuable or not, in the history of ideas has been buttressed by a comparable mix of forces, and in this sense the deconstructionists and the cultural relativists—so often the subject of mockery lately—have it right to a certain extent, though they tend to place undue stress on purely ideological forces. (Miasma was as much a creature of biology as of politics.) The river of intellectual progress is not defined purely by the steady flow of good ideas begetting better ones; it follows the topography that has been carved out for it by external factors. Sometimes that topography throws up so many barricades that the river backs up for a while. Such was the case with miasma in the mid–nineteenth century.

But most of these dams eventually burst. Yes, the path of science works within regimes of agreement and convention, and history is littered with past regimes that were overthrown. But some regimes are better than others, and the general tendency in science is for explanatory models to be overthrown in the name of better models. Oftentimes because their success sows the seeds of their destruction. Miasma became so powerful that it inspired a massive, state-sponsored intervention in the daily lives of millions of people, clearing the air by draining the cesspools. That intervention, miscalculated as it was, had the paradoxical effect of making the patterns of the epidemic more visible, at least to eyes that were capable of seeing them. And seeing the patterns more clearly means progress, in the long run at least.

JOHN SNOW SPENT MOST OF TUESDAY SEARCHING FOR patterns. In the morning he was knocking on doors, interrogating strangers in the street, asking anyone he encountered for anecdotal evidence about the outbreak and its victims. The clues he found were tantalizing, but too many doors went unanswered, and the dead couldn't report on their recent drinking habits. Personal testimony would not take him far in an evacuation zone. And so at midday he paid a visit to the Registrar-General's Office, where Farr gave him an early look at the numbers being calculated for the week. Eighty-three deaths had been reported in Soho between Thursday and Saturday. Snow asked for a complete list, including addresses, and returned to Broad Street to continue his sleuthing. He stood at the base of the pump, and ran through the addresses on the list. From time to time, he gazed out at the empty streets around him, imagining the paths the residents might take to find their way to water.

It was going to take more than body counts to prove that the pump was the culprit behind the Broad Street epidemic. Snow was going to need footprints, too.

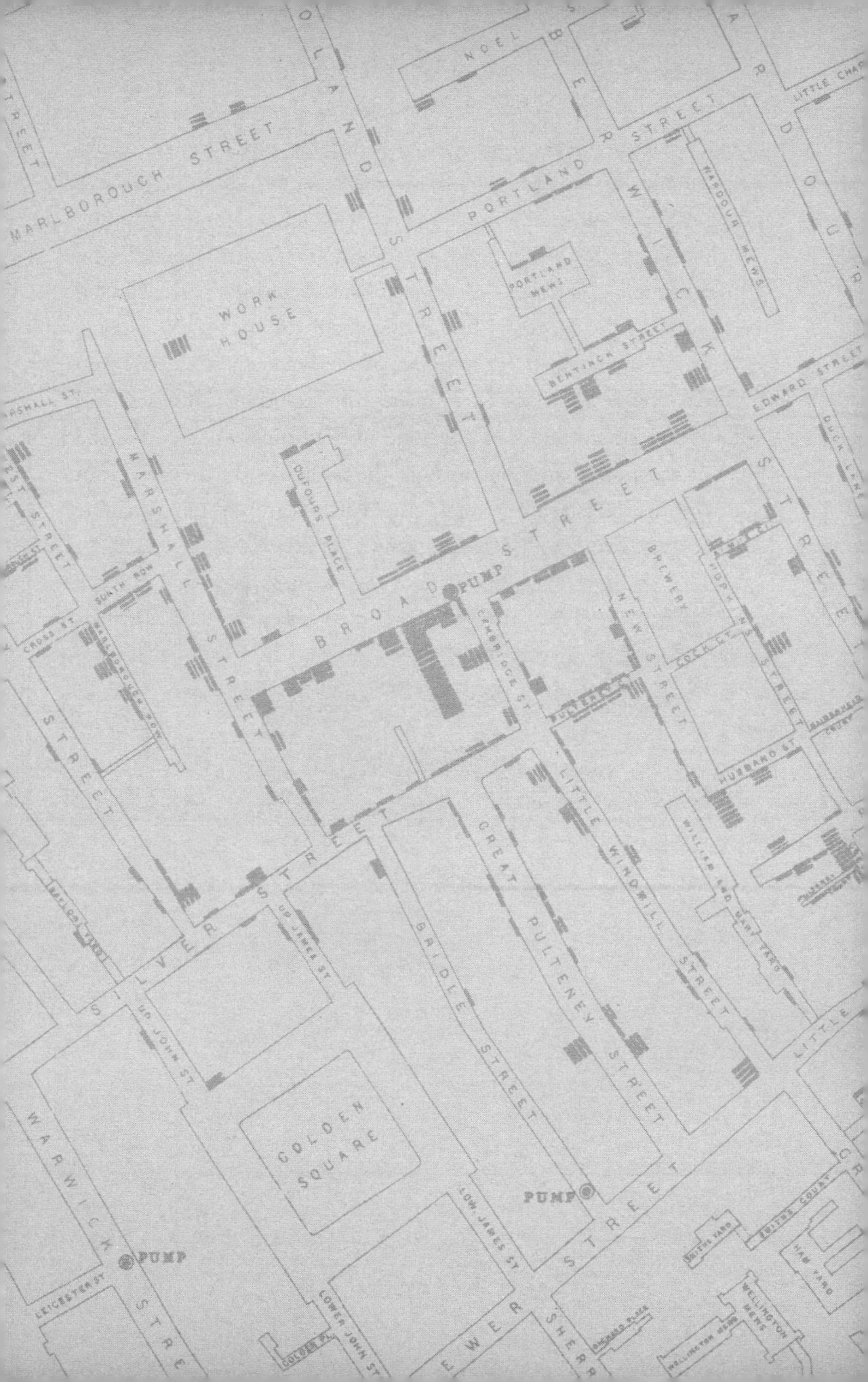

MARLBOROUGH STREET
NOEL
PORTLAND STREET
PORTLAND MEWS
WARDOUR MEWS
WORK HOUSE
BENTINCK STREET
EDWARD STREET
MARSHALL ST.
MARSHALL STREET
DUFOURS PLACE
BROAD STREET
PUMP
CAMBRIDGE ST
BREWERY
NEW STREET
HOPKINS STREET
CROSS ST
SOUTH ROW
MARLBOROUGH ROW
HUBBARD ST
GREAT PULTENEY STREET
LITTLE WINDMILL STREET
WILLIAM AND MARY YARD
BRIDLE STREET
SILVER STREET
UP JAMES ST
UP JOHN ST
GOLDEN SQUARE
WARWICK STREET
PUMP
LEICESTER ST
LOW. JAMES ST
LOWER JOHN ST
PUMP
STREET
SMITHS COURT
SMITHS YARD
WELLINGTON MEWS
HAM YARD

"BLUE STAGE OF THE SPASMODIC CHOLERA"

Wednesday, September 6

BUILDING THE CASE

A HUNDRED YARDS WEST OF THE BROAD STREET PUMP, IN the dark alley of Cross Street, a tailor lived in a single room at number 10, sharing the space with his five children, two of whom were fully grown. On warm summer nights the heat in their cramped living space could be unbearable, and the father would often wake after midnight and send one of the boys out to fetch some cool well water to combat the sweltering air. They lived only two blocks from the pump at Little Marlborough Street, but that water had such an offensive smell that they regularly walked the extra block to Broad Street.

The tailor and his twelve-year-old boy had been struck in the first hours of the outbreak, and both were dead by Saturday. Snow had found their address listed in the inventory of deaths that Farr had supplied him. Several other deaths were recorded on Cross Street as well. The location had caught Snow's eye when he first arrived back at the pump to survey the surrounding streets, armed with the addresses of the dead. Almost half the deaths Farr had recorded were linked to addresses within his line of sight; and half the remaining ones came from residences that were only a matter of steps from Broad Street itself. The Cross Street deaths were unusual, though: to make it to the Broad Street pump from there, you had to wind your way through two small side streets, then take a right onto Marshall Street, then another left, and then walk a long block down Broad Street. To get to the Little Marlborough pump, though, you simply strolled down the alley, walked two short blocks north, and you were there. It was within your line of sight if you stood at the very end of Cross Street.

Snow had noticed another element while scanning Farr's records: the deaths on Cross Street were much less evenly distributed than the ones in the immediate vicinity of the pump. Almost every house along Broad Street had suffered a loss, but there were only a handful of isolated cases on Cross Street. This is what Snow was looking for now. He could see at a glance that he'd be able to demonstrate that the outbreak was clustered around the pump, yet he knew from experience that that kind of evidence, on its own, would not satisfy a miasmatist. The cluster could just as easily reflect some pocket of poisoned air that had settled over that part of Soho, something emanating from the gulley holes or cesspools—or perhaps even from the pump itself. Snow knew that the case would be made in the exceptions to the rule. What he needed now were aberrations, deviations from the norm. Pockets of life where you would expect death, pockets of death where you would expect life. Cross Street was closer to Little Marlborough, and thus should have been spared in the outbreak, according to Snow's theory. And indeed, it had largely been spared, but for the four cases Farr had reported. Could those cases have some connection to Broad Street?

Sadly, by the time Snow arrived at 10 Cross to interview the tailor's surviving children, he was too late. He learned from a neighbor that the entire family—five children and their father—had died in the space of four days. Their late-night thirst for Broad Street water had destroyed them all.

IN HIS MIND SNOW WAS ALREADY DRAWING MAPS. HE'D imagined an overview of the Golden Square neighborhood, with a boundary line running an erratic circle around the Broad Street pump. Every person inside that border lived closer to the poisoned well; everyone outside would have had reason to draw water from a different source. Snow's survey of the neighborhood, based on Farr's initial data, revealed ten deaths that lay outside the boundary line. Two of them were the tailor and his son on Cross Street. After a few hours of conversation, Snow determined that three others were children who went to school near Broad Street; their grieving parents reported that the children had often drunk from the pump on their way to and from school. Relatives confirmed that three other casualties had maintained a regular habit of drawing water from Broad Street, despite living closer to another source. That left two remaining deaths outside the border with no connection to Broad Street, but Snow knew that two cholera deaths over a weekend was well within the average for a London neighborhood at that time. They might easily have contracted the disease from a different source altogether.

Snow knew that his case would also revolve around the inverse situation: residents who lived near the pump who survived, because, for one reason or another, they had opted not to drink from the poisoned well. He reviewed Farr's list again, looking this time for telltale absences. There were a handful of deaths reported at 50 Poland Street. On its own, this was a predictable number: Poland Street lay immediately to the north of the pump, well within Snow's imagined border. But in scanning the list, Snow realized that the number was strikingly low, because 50 Poland Street was the address of the St. James Workhouse, home to 535 people. Two deaths was routine for a household of ten living off of Broad Street. A population of five hundred living close to the pump should have seen dozens of death. As Whitehead had already learned from his daily rounds, the workhouse—despite its destitute and morally suspect inmates—had been something of a sanctuary from the outbreak. When Snow interrogated the workhouse directors, an explanation immediately jumped out at him: the workhouse had a private supply from the Grand Junction Water Works, which Snow knew from his earlier research to be one of the more reliable sources of piped water. The workhouse also had its own well on the premises. They had no reason to venture out to the Broad Street pump for water, even though it lay not fifty yards from their front door.

Snow noticed another telling absence on Farr's list. With seventy workers, the Lion Brewery at 50 Broad was the second-largest employer in the immediate vicinity. Yet not a single death was recorded for that address in Farr's list. It was possible, of course, that the workers had gone home to die, and so Snow paid a visit to the Lion's proprietors, Edward and John Huggins, who reported with some bafflement that the plague had passed over their establishment. Two workers had reported mild cases of diarrhea, but not a single one had shown severe symptoms. When Snow inquired about the water supply on the premises, the Hugginses replied that, like the workhouse, the brewery had both a private pipeline and a well. But, they explained for the benefit of the teetotaling doctor, they rarely saw their men drink water at all. Their daily rations of malt liquor usually satisfied their thirst.

Later, Snow would visit the Eley Brothers factory, where he found the situation much more dire. The proprietors reported that dozens of their employees had fallen ill, many of them dying in their own homes over the first few days of the epidemic. When Snow noticed the two large tubs of water that the brothers kept on premises for their employees to drink from, he scarcely needed to ask where the water had originated.

Snow had heard through the grapevine that the Eley brothers' mother and their cousin had recently perished of cholera as well, both of them far removed from Golden Square. The coincidence must have immediately struck Snow; perhaps he even thought back to the *experimentum crucis* gauntlet thrown down by the *London Medical Gazette* so many years before. No doubt, considering Snow's discretion, he posed the question delicately: Had Susannah Eley by any chance consumed some of the water from the Broad Street pump? It must have been an anguished moment for Snow: how to extract the information he needed without revealing that the brothers' thoughtfulness had been the agent of their mother's demise. Snow's taciturn demeanor would have helped him as the brothers described their regular deliveries of pump water to Hampstead; a more volatile investigator might have responded to the revelation of that crucial clue with more emotion. But whatever emotion he showed the Eley brothers, when he stepped out of the factory into the bright light of Broad Street, he must have thought to himself with some satisfaction that the case was coming together quite nicely indeed. The miasmatists might finally have met their match.

THERE IS A KIND OF MYTHOLOGY THAT STORIES LIKE THIS one tend inevitably to drift toward: the lone genius shaking off the chains of conventional wisdom through the sheer force of his intellect. But in explaining Snow's battle against the miasma theory and the medical establishment, it's not sufficient to point to his brilliance or his tenacity alone, though no doubt those characteristics played a crucial role. If the dominance of the miasma model was itself shaped by multiple intersecting forces, so, too, was Snow's ability to see it for the illusion that it was. Miasma was the intellectual equivalent of a contagious disease; it had spread through the intelligentsia with an alarming infection rate. So why was John Snow immune?

Part of the answer lies in Snow's study of ether and chloroform. The underlying insight that brought him his first round of acclaim was that the vapors of ether and chloroform had remarkably predictable effects on human beings. If you controlled the density of the gas, there was very little variation in the way humans—not to mention the frogs and birds in Snow's lab—would respond to inhalation. Without that predictability, of course, Snow would have never been able to build a thriving career for himself as an anesthesiologist; the risks and unreliability of the procedure would have greatly outweighed the benefits. Ether was itself a poisonous vapor—a kind of miasma in its own right—and yet it seemed entirely indifferent to the "inner constitution" of the humans who inhaled it. If ether had followed the pattern described by some of the miasmatists, it would have triggered radically different responses, depending on the inner constitution of each patient—perhaps causing some to become preternaturally alert, while inducing laughter in others, and rendering others senseless in seconds. But Snow had watched thousands of patients be sedated by the gas over the preceding six years, and he knew firsthand how mechanistic the process was. His whole career was, in a sense, a testimony to the predictable physiological effects of inhaled vapors. And so, when the miasma theorists invoked the inner constitution to explain why half the population of a room might succumb to poisonous vapors while the other half emerged unscathed, Snow was naturally inclined to view the theory with some suspicion.

His experience with choloform and ether had also endowed Snow with an intuitive grasp of the way that gases disperse in the environment. Ether could be deadly in a concentrated form, delivered directly to the patient's lungs. But a doctor delivering it, standing a foot away from the patient, wouldn't feel its effects in the slightest, because the density of ether molecules in the air dropped at a precipitous rate the farther removed one was from the inhaler itself. This principle—known as the law of diffusion of gases—had already been discovered and analyzed by the Scottish chemist Thomas Graham. Snow brought the same logic to miasma: if there were poisonous elements floating in the air, emanating from the cesspools or the bone-boilers, they were likely to be so massively dispersed that they posed no health risk. (Snow was only half-right on this point, of course: the vapors proved irrelevant where epidemic disease was concerned, but they did in fact have long-term deleterious effects, in that many of the industrial fumes of the age were carcinogens.) Several years after the Broad Street epidemic, Snow would make this connection explicit, in a controversial appearance before one of Benjamin Hall's public-health committees, defending the "offensive trades" (the bone-boilers, soap and dye makers, gut spinners) that stood accused of poisoning London's air. "I have arrived at the conclusion," Snow explained to the scandalized committee, "[that the offensive trades] are not injurious to the public health. I consider that if they were injurious to the public health they would be extremely so to the workmen engaged in those trades, and as far as I have been able to learn, that is not the case; and from the law of the diffusion of gases, it follows, that if they are not injurious to those actually upon the spot, where the trades are carried on, it is impossible they should be to persons further removed from the spot." Call it the Sewer-Hunter Principle: if all smell truly was disease, then a scavenger descending into an underground tunnel of raw waste should be dead in seconds.

Snow was also a doctor, a trained observer of physical symptoms, and he understood that the bodily effects of a disease were likely to offer important clues about the disease's original cause. In the case of cholera, by far and away the most pronounced change in the body lay in the small intestine. The disease invariably began with that terrible expulsion of fluids and fecal matter; all the other symptoms followed from that initial loss of water. Snow couldn't say exactly what kind of element was behind cholera's catastrophic attack on the human body, but he knew from observation that it invariably launched that attack from one place: the gut. The respiratory system, on the other hand, was largely unaffected by cholera's ravages. For Snow, that suggested an obvious etiology: cholera was ingested, not inhaled.

Snow's observational talents extended beyond the human body. The sad irony of his argument for the waterborne theory of cholera is that he had all the primary medical explanations in place by the winter of 1848–1849, and yet they fell on deaf ears for almost a decade. The tide eventually turned not because of his skills as a doctor or scientist. It wasn't lab research that would ultimately persuade the authorities; it wasn't direct observation of *V. cholerae* itself. It was Snow's faithful, probing observation of urban life and its everyday patterns: the malt-liquor drinkers at the Lion Brewery; the late-night trips for cold water on hot summer nights; the tangled web of private water supplies in South London. Snow's breakthroughs in anesthesia had revolved around his polymath skills as a physician, researcher, and inventor. But his cholera theory would ultimately depend on his skills as a sociologist.

Equally important was the social connection Snow had to the subjects he observed. It is not an accident that of the dozens and dozens of cholera outbreaks that he analyzed in his career, the one for which he is most famous erupted six blocks from his residence. Like Henry Whitehead, Snow brought genuine local knowledge to the Broad Street case. When Benjamin Hall and his public-health committee made their triumphant appearance on the streets of Soho, they were little more than tourists, goggling at all the despair and death, and then retreating back to the safety of Westminster or Kensington. But Snow was a true native. That gave him both an awareness of how the neighborhood actually worked, and it gave him a credibility with the residents, on whose intimate knowledge of the outbreak Snow's inquiry depended.

Snow shared more than geography with the working poor of Golden Square, of course. While he had long since elevated himself in social status, his roots as the son of a rural laborer colored his perception of the world throughout his life—primarily in the sense of blocking out certain dominant ideas. Nowhere in Snow's writings on disease does one ever encounter the idea of a moral component to illness. Equally absent is the premise that the poor are somehow more vulnerable to disease thanks to some defect in their inner constitution. Ever since he observed the Killingsworth mining outbreak as a young apprentice, Snow had long known that epidemics tended to afflict the lower orders of society. For whatever reason—probably some mix of rational observation and his own social awareness—that disparity led Snow to seek external causes, not internal ones. The poor were dying in disproportionate numbers not because they suffered from moral failings. They were dying because they were being poisoned.

Snow's resistance to the miasma theory was methodological as well. The strength of his model derived from its ability to use observed phenomena on one scale to make predictions about behavior on other scales up and down the chain. Observed failure of certain organ systems of the body could predict behavior of the whole person, which could in turn predict behavior in the social body en masse. If the symptoms of the cholera concentrated around the small intestine, then there must be some telltale characteristic in the eating and drinking habits of cholera victims. If cholera was waterborne, then the patterns of infection must correlate with the patterns of water distribution in London's neighborhoods. Snow's theory was like a ladder; each individual rung was impressive enough, but the power of it lay in ascending from bottom to top, from the membrane of the small intestine all the way up to the city itself.

And so Snow's immunity to the miasma theory was as overdetermined as the theory itself. Partly it was an accident of professional interest; partly it was a reflection of his social consciousness; partly it was his consilient, polymath way of making sense of the world. He was brilliant, no doubt, but one needed only to look to William Farr to see how easily brilliant minds could be drawn into error by orthodoxy and prejudice. Like all those ill-fated souls dying on Broad Street, Snow's insight lay at the intersection point of a series of social and historical vectors. However brilliant Snow was, he would never have proved his theory—and might well have failed to concoct it in the first place—without the population densities of industrial London, or Farr's numerical rigor, or his own working-class upbringing. This is how great intellectual breakthroughs usually happen in practice. It is rarely the isolated genius having a eureka moment alone in the lab. Nor is it merely a question of building on precedent, of standing on the shoulders of giants, in Newton's famous phrase. Great breakthroughs are closer to what happens in a flood plain: a dozen separate tributaries converge, and the rising waters lift the genius high enough that he or she can see around the conceptual obstructions of the age.

You can see the convergence of all these forces in Snow's regimen that Wednesday. In the midst of the most important investigation of his life, he was still a working physician, managing the diffusion of gases. He delivered chloroform to two patients: one having hemorrhoids removed, the other having a tooth extracted. He spent the rest of the day in the streets of his neighborhood, probing, questioning, listening. Yet each conversation, however intimate, was shaped by the impersonal calculations of Farr's statistics. He drew lines of connection between individual pathology and the wider neighborhood; he shifted perspective seamlessly from doctor to sociologist to statistician. He drew maps in his head, looking for patterns, looking for clues.

HENRY WHITEHEAD DIDN'T POSSESS A THEORY OF CHOLERA of his own, but he'd been steadily knocking down other ones for days now. He knew that the well-to-do neighborhoods around Golden Square were abuzz with sneering explanations for the outbreak: the poor of Soho, on the mean side of Regent Street, had brought this upon themselves. Either their physical crisis was the embodiment of a moral crisis, a kind of divine retribution, or they had succumbed to the fear of disease, which in turn made the cholera more powerful over them. Whitehead had been stewing over these calumnies for days now, but his outrage reached a fever pitch when James Richardson, St. Luke's scripture reader, failed to report for the noon vestry meeting. Richardson was one of Whitehead's closest friends, a blustery former grenadier guard with a fondness for debating metaphysics late into the evening. Whitehead found him at his home, suffering from a cholera attack that had begun several hours before. Richardson recounted a conversation he had had with a frightened neighbor who had asked about the best antidote to ward off the cholera. "I don't know what to take, but I do know what to do. It may neither prevent nor cure cholera; but it will save me from what is worse than cholera, i.e. from fear. I shall look up to my God, and though he slay me, yet I will trust in him."

If James Richardson—the very image of courage—could contract the disease, Whitehead thought, then the "inner constitution" explanation must certainly be false. With the number of new cases seemingly in decline, and with so much of the neighborhood emptied out, Whitehead finally had time to take stock of the situation, and he began thinking about ways to combat the popular prejudices. He was not a man of science, of course, but he knew as much about the path of the outbreak as did anyone, and perhaps if he wrote down his experiences, they would prove to be of some value to the wider population. Farr's *Weekly Returns,* published in that morning's *Times,* included this understated line: "On the north side of the Thames there has been a remarkable outbreak in the St. James District." The abrupt nature of the description was almost an insult. The true story of the Golden Square outbreak had yet to be told.

Richardson had mentioned one thing in passing that stuck with Whitehead as he returned to St. Luke's. The scripture reader had drunk a glass of water from the Broad Street pump on Saturday, a day or two before his symptoms appeared. It was not his habit to drink from the pump, and Richardson wondered if perhaps there had been a connection between that glass and his subsequent illness. But Whitehead thought the connection unlikely. He had personally seen so many residents recover from cholera after drinking Broad Street water. He himself had enjoyed a glass a few nights before, and had thus far resisted the plague. Perhaps Richardson had drunk too little.

WHAT WAS HAPPENING BELOW GROUND, IN THE DANK waters of the Broad Street well? The truth is, we don't know. Clearly by Wednesday, it was significantly harder for the *V. cholerae* to find its way to a human small intestine, mostly because the number of people using the pump had dropped so precipitously, thanks to the death toll and the mass exodus. In that sense, it's possible that the *V. cholerae*'s dramatic reproductive success over the weekend—think how many trillions of bacteria had been created in that short amount of time—had been the agent of its own demise. Establishing a base in a popular watering hole in London's most densely populated district allowed the bacteria to spread through the neighborhood like wildfire, but the fire's spread was so sudden and so extensive that it quickly burned through its primary fuel supply. There weren't enough small intestines left to colonize.

It's also possible that the *Vibrio cholerae* had not been able to survive more than a few days in the well water below the Broad Street pump. With no sunlight penetrating the well, the water would have been free of plankton, and so the bacteria that didn't escape might have slowly starved to death in the dark, twenty feet below street level. The purity of the well water may have played a role as well.

V. cholerae much prefers water with a high saline content, or with extensive organic material. In distilled water, the organism dies off in a matter of hours. But the most likely scenario is that the bacterium was itself in a life-or-death struggle with another organism: a viral phage that exploits *V. cholerae* for its own reproductive ends the way *V. cholerae* exploits the human small intestine. One phage injected into a bacterial cell yields about a hundred new viral particles, and kills off the bacterium in the process. After several days of that replication, the population of *V. cholerae* might have been replaced by phages that were harmless to humans.

Whatever the explanation, those few days at the very outset of the epidemic had been a kind of microbial lottery: a population of *V. cholerae* gathered together in a small pool of water, waiting to be propelled upward into the light of day, where untold possibilities for reproductive glory awaited them. Those that made it out of the pump would go on to generate trillions of offspring in the small intestines of their victims. Those that stayed behind would die.

When Whitehead later retraced the week's events, he found even more cases of survivors drinking copious amounts of Broad Street water. He tracked down one boy who had fallen ill and attributed his recovery to his drinking ten quarts; he found a girl who consumed seventeen during her (ultimately successful) attempt to fight off the disease. But he found something else as well in re-creating the outbreak's chronology: almost all the survivors who had consumed large quantities of Broad Street water did their drinking after Saturday. It was much harder to find anyone who would report drinking the pump water earlier in the week—because most of those people were dead.

So it is possible that *V. cholerae* had largely abandoned the pump by the weekend, dying in the dark, cool waters as the outbreak flamed twenty feet above. Perhaps another microbial organism had vanquished the killer on its own. Or perhaps the natural flow of groundwater had slowly cleansed the pump supply and the initial colony of *V. cholerae* had dispersed through the gravel and sand and clay beneath the streets of Soho.

BY THE END OF THE DAY, SNOW HAD BUILT A CONVINCING statistical case against the pump. Of the eighty-three deaths recorded on Farr's list, seventy-three were in houses that were closer to the Broad Street pump than to any other public water source. Of those seventy-three, Snow had learned, sixty-one were habitual drinkers of Broad Street water. Only six of the dead were definitively not Broad Street drinkers. The final six remained mysteries, "owing to the death or departure of every one connected with the deceased individuals," as Snow would later write. The ten cases that fell outside the imagined boundary line surrounding the Broad Street pump were equally telling: eight appeared to have a connection to Broad Street. Snow had established new causal chains back to the pump water, beyond the list of Farr's addresses: the proprietor of the coffeehouse who often sold sherbet mixed with Broad Street water told Snow that nine of her customers had died since the outbreak began. He had drawn the telling contrast between the Lion Brewery and the Eley Brothers factory; he had documented the unlikely safe haven of the Poland Street Workhouse. He even had his *experimentum crucis* in Hampstead.

It was, on the face of it, a staggering display of investigative work, given the manic condition of the neighborhood itself. In the twenty-four hours since he'd received Farr's early numbers, Snow had tracked down intimate details of behavior from the surviving family and neighbors of more than seventy people. The fearlessness of the act still astonishes: as the neighborhood emptied in terror from the most savage outbreak in the city's history, Snow spent hour after hour visiting the houses that had suffered the worst—houses that were, in fact, still under assault. His friend and biographer Benjamin Ward Richardson later recalled: "No one but those who knew him intimately can conceive how he laboured, at what cost, and at what risk. Wherever cholera was visitant, there was he in the midst."

It's unlikely that anyone in London that day had a better sense of the outbreak's magnitude than John Snow and Henry Whitehead. But ironically, their local knowledge of Broad Street made it hard for them to gauge the true extent of the tragedy. There were at least twice as many Soho residents suffering in a local hospital as there were people dying in the shuttered dark of their own homes. In the three days after September 1, more than 120 cholera patients overwhelmed the staff at nearby Middlesex Hospital, where Florence Nightingale observed that a disproportionate number of the sufferers appeared to be prostitutes. The sick were stacked together in large, open rooms, and treated with saline and calomel. The air was thick with the smell of chlorine and sulfuric acid that the staff had scattered around the sickrooms in large dishes, ostensibly to purify the air. It was ultimately of little use: two-thirds of the patients died.

When the number became too great to house at Middlesex, new arrivals were sent to the University College Hospital. Twenty-five cholera patients arrived in the first three days of September. Westminster Hospital had admitted eighty patients in the first few days of the outbreak. Other institutions saw notable influxes: by that Wednesday, more than fifty cholera patients had been admitted to Guy's, St. Thomas', and Charing Cross hospitals.

St. Bartholomew's Hospital had received the most cholera patients—almost two hundred in the first days of the outbreak. The physicians there experimented with multiple treatments, with varying degrees of success: castor oil, capsicum, even cold water. An intravenous injection of a saline solution designed to mirror the salinity of blood serum appeared to revive two patients, but they died hours later—most likely because, like Thomas Latta's patients in 1832, they were not given multiple injections.

And so the devastation in the streets above Golden Square was, in truth, only a fraction of the story. As Snow and Whitehead made their calculations that Wednesday, they were still thinking in double-digit figures. They would soon discover that those numbers were shockingly optimistic.

IT IS POSSIBLE SNOW'S INTENSE ROUND OF QUESTIONING may have, on its own, diminished the spread of the epidemic. We know from Snow's own account that he spoke with hundreds of people in the neighborhood over the course of the week, and that most of those conversations involved questions about the Broad Street pump. What we don't know is whether Snow betrayed his theory of the cholera's source in those conversations. Were they both interviews *and* warnings? Snow was a physician, after all, and the poor, frightened inhabitants of Soho were his patients. If he believed that the pump was spreading fatal disease, it seems unlikely that he would have kept that information to himself. Would a hundred separate warnings from an esteemed physician be enough to suppress the neighborhood's taste for Broad Street water? The most dramatic drop in deaths had occurred on Tuesday and Wednesday—two days after Snow began exploring the neighborhood. Perhaps fewer people were dying because some portion of the population had heard a rumor that the pump was to blame.

But if the epidemic was in decline, it was still at terrifying levels by any normal standard. Snow knew from his investigative rounds that at least a dozen new deaths had occurred that Wednesday—ten times the normal rate for the neighborhood. Given the exodus of the population, it was possible that the plague was still every bit as deadly, on a per capita basis. He knew his statistical account of the outbreak would be a convincing argument for his waterborne theory, particularly when it was accompanied by the final results from his South London waterworks study. His monograph on cholera would have to be revised, new articles submitted to *The Lancet* and the *London Medical Gazette.* But in the short term, there was a more pressing matter at hand. People were still dying in his neighborhood, and his survey of the outbreak clearly revealed the culprit.

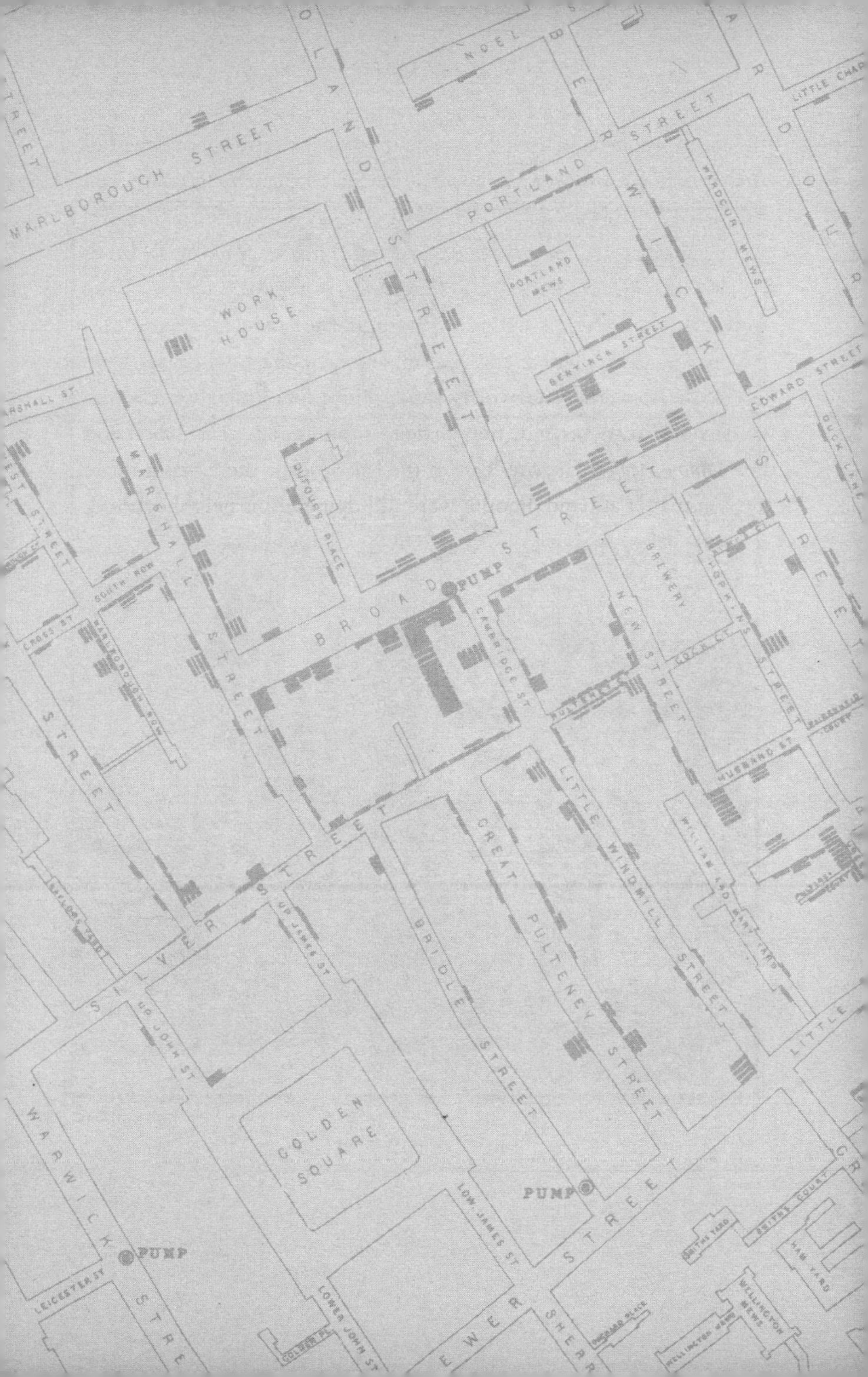

MARLBOROUGH STREET
POLAND STREET
NOEL
BERWICK STREET
PORTLAND STREET
LITTLE CHAP
WARDOUR STREET
WARDOUR MEWS
WORK HOUSE
PORTLAND MEWS
BENTINCK STREET
EDWARD STREET
DUCK LANE
MARSHALL ST
MARSHALL STREET
DUFOURS PLACE
BROAD STREET
PUMP
CAMBRIDGE ST
BREWERY
NEW STREET
HOPKINS STREET
CROSS ST
SOUTH ROW
MARLBOROUGH ROW
HUSBAND ST
LITTLE WINDMILL STREET
GREAT PULTENEY STREET
WILLIAM AND MARY YARD
SILVER STREET
UP JAMES ST
UP JOHN ST
BRIDLE STREET
GOLDEN SQUARE
WARWICK STREET
PUMP
LOW JAMES ST
PUMP
LEICESTER ST
LOWER JOHN ST
GOLDEN PL
SMITHS YARD
SMITHS COURT
HAM YARD
WELLINGTON MEWS

"DEATH'S DISPENSARY"

Friday, September 8

THE PUMP HANDLE

On Thursday night, the Board of Governors of St. James Parish had held an emergency meeting to discuss the ongoing outbreak and the neighborhood's response. Halfway into the meeting, they received notice that a gentleman wished to address them. It was John Snow, armed with his survey of the past week's devastation. He stood before them, and in his odd, husky voice told them that he knew the cause of the outbreak, and could prove convincingly that the great majority of cases in the neighborhood could be traced to its original source. It is unlikely that Snow went into the intricacies of his broader case against the miasma theory—better to go straight to the telling patterns of death and life, leave the philosophizing for another day. He explained the dismal ratios of survival among the people living near the pump, and the unusual exemptions granted to people who had not drunk the water. He told the Board of Governors of deaths that had transpired far from Golden Square, connected to the area only by the consumption of Broad Street water. He may have told them of the brewery of the workhouse on Poland Street. Death after death after death had been linked to the water at the base of the Broad Street well. And yet the pump remained in active use.

The members of the Board were skeptical. They knew as well as any other locals how highly regarded the Broad Street water was—particularly as compared to the other nearby pumps. But they also knew firsthand the smells and noxious fumes that were rampant in the neighborhood; surely these were more responsible for the outbreak than the reliable Broad Street water. Yet Snow's argument was persuasive—and, besides, they had few other options. If Snow was wrong, the neighborhood might go thirsty for a few weeks. If he was right, who knew how many lives they might save? And so, after a quick internal consultation, the Board voted that the Broad Street well should be closed down.

The following morning, Friday, September 8, exactly a week after the outbreak had first begun its awful rampage through Soho, the pump handle was removed. Whatever menace lay at the bottom of the well would stay there for the time being.

The deaths in Soho would continue for still another week, and the final reckoning of the assault of the Broad Street well on the neighborhood would not be calculated for months. The removal of the pump handle was generally ignored by the newspapers. On Friday, the *Globe* had published an upbeat—and typically miasmatic—account of the present state of the neighborhood: "Owing to the favourable change in the weather, the pestilence which has raged with such frightful severity in this district has abated, and it may be hoped that the inhabitants have seen the worst of the visitation. Yesterday there were very few deaths, and this morning no new cases were reported." On the following day, however, the news appeared to be less encouraging:

> We regret to announce that after the account was written which appeared in The Globe of yesterday, there were several severe and fatal cases of cholera, and that seven or eight were reported on Saturday morning, although the wisest precautions were adopted to arrest the progress of the disease. The neighbourhood of Golden-square presented . . . a most melancholy and heart-rending appearance. There was scarcely a street free from hearses and mourning coaches, and the inhabitants of the district, appalled by the calamity which has visited them, crowded the streets to witness the last sorrowing act of duty towards their neighbours and friends. A vast number of the tradespeople left their shops and fled from the place, the closed shutters bearing the announcement that business had been suspended for a few days. Messers Huggins, the brewers, with praiseworthy forethought, have issued an announcement that the poor . . . may obtain any quantity of hot water for cleansing their dwellings, or other purposes, at any hour of the day or night, an act of humanity and kindness of which a large number have availed themselves.

Dozens would die over the next week, but clearly the worst was over. When the final numbers were tallied, the severity of the outbreak shocked even those who had lived through it. Nearly seven hundred people living within 250 yards of the Broad Street pump had died in a period of less than two weeks. Broad Street's population had literally been decimated: ninety out of 896 residents had perished. Among the forty-five houses extending in all directions from the intersection of Broad and Cambridge streets, only four managed to survive the epidemic without losing a single inhabitant. "Such a mortality in so short a time is almost unparalleled in this country," the *Observer* noted. Past epidemics had produced higher body counts citywide, but none had killed so many in such a small area with such devastating speed.

THE REMOVAL OF THE PUMP HANDLE WAS A HISTORICAL turning point, and not just because it marked the end of London's most explosive outbreak. History has its epic thresholds where the world is transformed in a matter of minutes—a leader is assassinated, a volcano erupts, a constitution is ratified. But there are other, smaller, turning points that are no less important. A hundred disparate historical trends converge on a single, modest act—some unknown person unscrews the handle of a pump on a side street in a bustling city—and in the years and decades that follow, a thousand changes ripple out from that simple act. It's not that the world is changed instantly; the change itself takes many years to become visible. But the change is no less momentous for its quiet evolution.

And so it was with the Broad Street well that the *decision* to remove the pump handle turned out to be more significant than the short-term effects of that decision. Yes, the Broad Street outbreak would burn itself out over the next few days, as the last victims died off and other, more fortunate, cases recovered. Yes, the neighborhood would slowly return to normalcy in the weeks and months that followed. These were real achievements that arose from that pump handle being removed, even if the water in the well had potentially been purged of *V. cholerae* by the time Snow made his case to the Board of Governors. But the pump handle stands for more than that local redemption. It marks a turning point in the battle between urban man and *Vibrio cholerae,* because for the first time a public institution had made an informed intervention into a cholera outbreak based on a scientifically sound theory of the disease. The decision to remove the handle was not based on meteorological charts or social prejudice or watered-down medieval humorology; it was based on a methodical survey of the actual social patterns of the epidemic, confirming predictions put forward by an underlying theory of the disease's effect on the human body. It was based on information that the city's own organization had made visible. For the first time, the *V. cholerae*'s growing dominion over the city would be challenged by reason, not superstition.

But learning to listen to reason takes time, particularly among the general public of Broad Street, who had heard nothing *but* superstition from the authorities for as long as cholera had been in London. When the Board of Governors removed the handle on Friday morning, the act was met with open jeering and derision by the passersby who chanced to witness it. Their bafflement is not hard to understand. For many survivors, Broad Street water had been their primary medicine. And now the authorities were going to cut off the supply? Were they trying to wipe out the entire neighborhood?

It was not just the Soho locals who were deaf to Snow's reason. The very day that the local Board of Governors removed the handle, the president of the national Board of Health, Benjamin Hall, issued directives for the three-man committee he had formed to investigate the Broad Street outbreak. The inspectors were asked to perform a house-to-house survey through the entire neighborhood and report back on a long list of environmental conditions. It is worth quoting the directives in their entirety, since the list captures perfectly the miasmatic obsessions of the Board of Health:

> Structural peculiarities of the Streets as regards Ventilation.
>
> Nuisances, slaughter-houses, noxious trades, etc.

Smells in the streets and their source, gully grates, gutters, etc; whether the gully grates trapped, whether cases and deaths more numerous in houses near gully grates.

Smells in houses and their source, such smells worse during the night, or in the morning before the houses or shops were opened.

Whether the house had privy, or water-closet or cesspool and the position of these; whether complaints of smells from them; whether they were in good condition; whether the water-closets were well supplied with water; whether the house drainage stopped. . . . This district has been lately drained. Ascertain how many of the houses have drains connected with the new sewers; whether the house drains pass under the house to reach the sewer; the structure of the house drains, pipe or brick drains, and their condition; whether subject to stoppage, or smells from them.

Examine the basements as to the depth of the floor below the level of the street; whether there had been any accumulations of house refuse in these basements, or in the adjoining cellars before the outbreak. Consider the effect of these conditions on the general ventilation of the house, especially at night. . . .

Examine the houses as to their general cleanliness and means of ventilation. Examine also the back yards, and inquire what was their condition before the epidemic. Note if they be flagged or filthy.

Examine whether the disease occurred in the upper or lower flats. Get, if possible, the proportion of cases in the flats.

Estimate as closely as you can the condition of the inhabitants as to overcrowding, personal cleanliness, habits, diet, etc.

Get the number of cases in each house, and the number of deaths of persons who lived in each house.

Examine the water supply as to its source, quality, amount, whether drawn from pipes or water-butts, and the condition of the butts.

> Note the general condition of the streets and courts, and inquire what was the state of the cleansing before the outbreak.
>
> Examine whether the disturbance of the ground in making a sewer through the old burial ground in Little Marlborough Street, or the filtrations from it into the sewer, or the drainage of any nuisance into the general sewerage of the district had had any effect, or whether the sewers had accumulations in them that might have been injurious.

Hall's instructions for his cholera committee offer a brilliant case study in how dominant intellectual paradigms can make it more difficult for the truth to be established, even if the people involved are smart and attentive and methodical in their research. Hall's list is a kind of straitjacket for an eventual document. You can tell from just scanning the instructions what kind of document they will ultimately produce: a rich and impossibly detailed inventory of the smells of Soho circa 1854. Half of the categories specifically mention smell and ventilation, and the few directives that might potentially be relevant to the waterborne theory of the disease—such as the condition of the cesspools—are specifically colored by Hall's concern about smell in each instance.

In all, Benjamin Hall delivered about fifty specific instructions to his committee. Only two of them—regarding the quality and source of the water supply—were essential to proving or disproving Snow's waterborne theory. But of course, on their own, those two variables were close to meaningless. Snow himself had detected nothing unusual in the water on Monday morning, at the height of the epidemic. Analyzing the quality of the water using the available technologies of the day couldn't shed light on the mystery either way: there was nothing to see. Pacini had caught a glimpse of the bacteria in his microscope that year, but he would be alone in his discovery for three decades. The most reliable way to "see" the cholera was indirectly, in the way the drinking habits of the neighborhood mapped onto the patterns of disease and death that Farr had captured in the *Weekly Returns.* If you failed to superimpose those two data sets, the power and clarity of the waterborne theory disappeared. But Hall never asked his committee to investigate the drinking habits of the population, much less compare those habits to the overall distribution of deaths.

It's crucial to note that Hall was not blind to the basic epidemiological principles that governed Snow's work—that the cause of a disease can be deduced by observing statistically unusual patterns in the course of an epidemic. Hall requested that the investigators report on whether the cholera deaths were concentrated around gully gates or the site of the plague burial ground. But the waterborne theory did not rise to that level of scrutiny. Despite the fact that Snow had published on the subject, and despite Snow's numerous conversations with William Farr about cholera and the water supply, the president of the Board of Health did not find it necessary to determine whether there was an unusual concentration of deaths around any of the neighborhood's sources of drinking water. Hall's instructions had rigged the game against Snow's theory from the very outset.

But Hall's task force would not be the only one investigating the Broad Street epidemic. In the weeks and months that followed the outbreak, another group would probe the neighborhood, piecing together the story, looking for clues. And at its center would be the one man who probably knew the neighborhood as intimately as anyone in Soho: Henry Whitehead.

News of the pump handle's removal had struck Whitehead as being particularly foolish. When he first heard the contaminated-pump theory that Friday, he reacted with a quick dismissal, siding with the jeering throngs on Broad Street. This will be easy enough to disprove, he thought. And Whitehead was uniquely equipped to do the disproving. Snow's two-day investigation couldn't compete with the bedside hours Whitehead had logged since the outbreak first erupted on Friday. The young curate had already been constructing arguments against other prevailing theories. Now he would add the waterborne theory to the list. The Board of Governers might have been easily swayed by Dr. Snow's demographic sleight of hand, but they didn't know the neighborhood as well as Whitehead did; they hadn't seen a girl drink seventeen quarts of pump water and survive. It would take some additional research, Whitehead knew, but he was confident that the pump would be exonerated in time.

"Every limit is a beginning as well as an ending," George Eliot would write a few years later in *Middlemarch*. So it is with the story of the pump handle's removal. It was the end of the Broad Street well's assault on Golden Square, and the beginning of a new era of public health. But it does not offer the easy closure of the detective story. The remaining residents did not gather around Dr. Snow to celebrate his solving the mystery of Broad Street; Benjamin Hall did not drop his miasma obsessions overnight; even the Board of Governers remained unimpressed with Snow's theory, though they followed his advice. And Henry Whitehead was so unconvinced by the case against the pump that he vowed to disprove it. So the true narrative arc of the Broad Street outbreak turns out to have a dialectical twist at its end:

in persuading the otherwise incompetent Board of Governers to follow his advice, Snow awakened the one adversary who possessed more local knowledge of the outbreak than himself. In overcoming one opponent, Snow created an even more daunting challenge for his waterborne theory. Snow still had a long list of potential converts to win over: Benjamin Hall and his miasma-addled investigators; William Farr; the editors of *The Lancet*. But in the short term, his primary nemesis would be the Reverend Henry Whitehead.

WHITEHEAD HAD BEEN INFORMALLY ASSEMBLING CLUES FROM the very outset. On that Friday, before receiving word of the pump handle's removal, he had ascended to the pulpit at St. Luke's to give the daily sermon. Standing in front of his haggard parishioners in the half-empty church, he noted the disproportionate number of poor, elderly women in the pews. He congratulated them on their "remarkable immunity from the pestilence." But even as he spoke the words, he wondered: How can this be? What kind of pestilence spares the old and the destitute?

In the months that followed, Whitehead and Snow explored Broad Street on separate but parallel tracks. Snow began integrating the data from his investigation into a new version of his cholera monograph from 1849, while writing a handful of articles for the medical journals that addressed the outbreak. The section of the monograph devoted to Broad Street began with these dramatic lines:

> The most terrible outbreak of cholera which ever occurred in this kingdom, is probably that which took place in Broad Street, Golden Square, and the adjoining streets, a few weeks ago. Within two hundred and fifty yards of the spot where Cambridge Street joins Broad

> Street, there were upwards of five hundred fatal attacks of cholera in ten days. The mortality in this limited area probably equals any that was ever caused in this country, even by the plague; and it was much more sudden, as the greater number of cases terminated in a few hours. The mortality would undoubtedly have been much greater had it not been for the flight of the population. Persons in furnished lodgings left first, then other lodgers went away, leaving their furniture to be sent for when they could meet with a place to put it in. Many houses were closed altogether, owing to the death of the proprietors; and, in a great number of instances, the tradesmen who remained had sent away their families: so that in less than six days from the commencement of the outbreak, the most afflicted streets were deserted by more than three-quarters of their inhabitants.

That fall, Whitehead quickly wrote and published a seventeen-page monograph titled *The Cholera in Berwick Street.* It was the first comprehensive look at the outbreak written for a general audience. Most of Whitehead's inquiries over those initial weeks were aimed at taking stock of the outbreak's reach and its duration. He began his monograph with a terse inventory:

> Dufour's Place. —Houses, 9; population 170; deaths, 9; houses without any deaths, 4. Rumour sadly exaggerated the mortality in this place.
>
> Cambridge Street. —Houses, 14; population, 179; deaths, 16; deaths on the west side, 10; east, 6, of which 3 were in one house. Five houses escaped.

Whitehead described the strange lack of connection he had observed at the height of the outbreak between the sanitary conditions and the mortality rates in each residence. He noted that the model home on Peter Street—the very one that had been commended by the authorities several years back for its cleanliness—had suffered twelve deaths, the largest number of any residence in the neighborhood. He traced the devastation that the outbreak had leveled against the neighborhood's families: "There were no less than 21 instances of husband and wife dying within a few days of each other. In one case, besides parents, 4 children also died. In another both parents and 3 of their 4 children. In another, a widow and 3 of her children." Not fifteen yards from the front steps of St. Luke's Church stood four houses that had, between them, lost thirty-three people.

Reading Whitehead's monograph, you can sense the young curate grappling with the theological implications of the outbreak. A visitation of plague had to be, on some level, an expression of divine will, and in this case the divinity appeared to have singled out the parish of St. Luke's for the most severe retribution imaginable. It must have been a vexing reality to face as a man of the cloth: of all the parishes in London, over the many years that cholera had ravaged the country, God had seen fit to subject Whitehead's own small community to the most explosive epidemic attack in the history of the city. In the monograph, Whitehead initially professes an inability to explain such an event in terms of divine will, but then he offers a half-formed theory, one that itself follows a markedly dialectical logic:

> God's ways are equal, man's ways are unequal; and another fact, less difficult to be accounted for, presents itself to our notice, even the unequal accumulation of filth and dirt, the overcrowding together of human beings, the culpable sufferance of ill-constructed streets and ill-ventilated houses, indifference to first principles of drainage and sewage, aggravating the pestilence in particular localities, but

> attracting little attention and exciting little alarm, till here and there a mine explodes, revealing to the startled population of an ill-managed city the peril of a position which admits of any one street or parish, and that none of the lowest and filthiest, becoming a huge charnel-house in a day or an hour.

Till here and there a mine explodes. The outbreak, as brutal as it was, nonetheless shed light on the poverty and despair of inner-city life, illuminating everyday suffering with the bright light of extraordinary despair. Whitehead had the story half right: the terrifying visibility of the outbreak did in fact sow the seeds of a cure. But it was not divine providence that drove the process. It was density. Crowd a thousand people into three city blocks and you create an environment where epidemic disease will flourish; but in flourishing, the disease reveals the telltale characteristics of its true nature. Its efflorescence points the way to its ultimate defeat. The Broad Street pump was a kind of urban antenna, sending out a signal through the surrounding neighborhood, a signal with a detectible pattern that allowed humans to "see" *V. cholerae* without the aid of microscopes. But without those thousand bodies crowded around the pump, the signal would have been lost, like a sound wave dissipating into silence in the vacuum of space.

In the weeks after the outbreak, Whitehead had observed enough of these patterns to debunk a number of prevailing theories in his monograph. His account of the devastation at Peter Street exposed the fallacy of the sanitary hypothesis; and he offered numerous cases of brave parishioners falling ill to combat the "fear kills" platitudes. He tabulated the ratio of deaths on upper and lower floors to demonstrate that the cholera had attacked both classes evenly. But despite his initial derision at the the pump handle's removal, the

Broad Street well goes unmentioned in the monograph. Perhaps Whitehead simply felt he hadn't accumulated enough evidence against Snow's case to include the waterborne theory in the text. Or perhaps his early inquiries had started to change his mind.

Either way, the monograph was only the beginning. Whitehead would end up pursuing details of the Broad Street outbreak further than he ever imagined in the coming months—further, indeed, than John Snow himself would venture. In late November, the vestry of St. James' voted to form a committee to investigate the Broad Street outbreak, initially planning to produce a report based on a questionnaire circulated through the neighborhood, augmented by the data assembled by the Board of Health committee. But when the vestry approached Benjamin Hall, the Board's president declined to share his committee's findings—"principally on the ground that investigations of this kind were more valuable when independent." The snub turned out to be fortuitous. Faced with limited returns from their questionnaire, and with no contribution from the Board of Health, the vestry recognized that they would have to assemble a team of their own investigators. On the merits of his recently published monograph, and recognizing the value of his knowledge of the community, they asked the Reverend Whitehead to join the committee. They also invited that local doctor who had been so agitated about the state of the Broad Street pump. Snow and Whitehead may not have agreed on the cause of the outbreak, but they were now working on the same team.

WHITEHEAD BEGAN HIS ASSAULT ON THE PUMP-CONTAMINATION theory by examining a crucial absence in Snow's original survey of the neighborhood. Snow had focused almost exclusively on the

Soho residents who had perished in the outbreak, detecting that an overwhelming majority of them had consumed Broad Street water before falling ill. But Snow had not investigated the drinking habits of the neighborhood residents who had *survived* the epidemic. If that group turned out to have drunk from the Broad Street pump at the same rate, then the whole basis for Snow's theory would dissolve. The connection between pump drinking and cholera would be meaningless if most of the neighborhood—the dead *and* the living—were drinking from the pump. Most of the dead had probably also strolled down Broad Street at some point in the days leading up to the epidemic, but that didn't mean that strolling down Broad Street caused cholera.

Whitehead's local knowledge gave him a crucial edge in this investigation, in that he was uniquely able to track down the hundreds of residents who had fled the neighborhood in the weeks after the outbreak. Snow would have intuitively understood the importance of surveying the pump-drinking rates among the survivors, but the great majority of the survivors were unreachable to him that first week of September. And so Snow had been forced to build his case against the pump on his survey of the dead, augmented by a few odd cases of unlikely survival (the workhouse, the brewery). Whitehead, on the other hand, could draw upon the extensive social network he'd long ago established in the neighborhood to track down the émigrés from Golden Square. His investigation took him throughout the Greater London area in the months that followed his appointment to the committee; when he learned of former residents who had moved outside the city, he sent inquiries by post. In the end, he tracked down information on 497 residents of Broad Street, more than half the population that had lived there in the weeks before the outbreak.

As he threw himself into the investigation, sometimes visiting the same flat five separate times to pursue new leads, Whitehead felt his resistance to the pump-contamination theory fade. Again and again, the recollections of cholera survivors would eventually turn up some forgotten connection to the Broad Street pump. A young widow whose husband died on the first had originally told Whitehead that the couple had not been Broad Street drinkers. But several days later, a memory came back to her: on the night of the thirtieth her husband had asked her to fetch some water from the pump to drink with dinner. She herself had not consumed any. One woman whose husband and daughter had come down with the cholera (eventually surviving it), denied forthrightly that anyone in the house had ever favored the Broad Street water. But when she relayed the details of her curious interview with the Reverend Whitehead to the rest of the family, the daughter recalled that, no, she had in fact drunk from the Broad Street well in the days before the outbreak.

This last case was typical of the stories that Whitehead unraveled: the children turned out to offer the missing link to the pump. In performing his analysis of the neighborhood's drinking customs, Whitehead noted how often the young people were asked to fetch water for their families. A visit to the Broad Street pump was a commonplace chore for any child over the age of six or seven, and their familiarity with the well meant that a number of neighborhood children had drunk from it without their parents' knowing. As he listened to these accounts, Whitehead's mind returned to the image of all those widows gathered at St. Luke's on the day the pump handle was removed. At last he had a potential explanation for their immunity. It wasn't that the ladies were somehow morally superior to the dead; it wasn't that they possessed sturdier constitutions or more hygienic lifestyles. What united them all as a group was that they were old and infirm and living alone, with the result that they didn't have anyone to fetch water for them.

As Whitehead tabulated his initial numbers, the case against the pump looked powerful indeed. Among the pump-water-drinking population, the rates of infection were along the lines that Snow had outlined in his original survey: for every two Broad Street drinkers who were not affected, there were three who fell ill. That ratio seemed even more striking when you compared it to the infection rates among those who had not drunk from the well: only one in ten of that group had been seized with the cholera. As much as he had resisted the waterborne theory, Whitehead found himself confronting the stubborn fact that choosing to drink from the well increased your odds of infection sevenfold.

Still, three objections to the pump-contamination theory continued to trouble Whitehead. Snow lived in Soho but he was not exactly a Broad Street regular, and Whitehead felt that his theory didn't square with the Broad Street well's long track record of supplying unusually pure water to the neighborhood. If a local watering hole was going to be infected with some kind of infectious agent, it was much more likely to be the foul-smelling supply at the Little Marlborough Street pump.

And then there were the survivors. The raw numbers made the case against the well look convincing, but Whitehead had a hard time shaking his firsthand observations: watching his parishioners drink gallons of Broad Street water from their seeming deathbeds—and then subsequently recover. Whitehead had his own survival in mind as well; he had drunk from the well at the very height of the outbreak, after all. If the well was truly poisoned, why had he been spared?

The course of the investigation had planted one additional objection in Whitehead's mind. In November, the Paving Board had undertaken an examination of the Broad Street pump, looking for some connection to the sewer lines that might have contaminated the well water with waste matter. Their verdict had been definitive: the investigators found the well "free from any fissures or other communication with drains or sewers by which such matters could possibly be conveyed into the waters." They also ran chemical and microscopy tests on the water itself, all of which failed to detect anything out of the ordinary.

John Snow's research would be critical in helping Whitehead find a way around his first objection, but it would be Whitehead himself who would solve the mystery posed by the other two. During these winter months, Snow had been revising his book on cholera, integrating both the data from his South London water supply survey and an account of the Broad Street epidemic. At some point early in 1855, he gave Whitehead a copy of the monograph. In reading through Snow's version of the previous September's events, the curate was surprised to find that Snow had not blamed the outbreak on a "general impurity in the water." Snow's theory had assumed that the original case was a "special contamination . . . from the evacuations of cholera patients" that had leaked into the well from a sewer or cesspool. So the *general* quality of the water wasn't relevant to Snow's theory. Whatever agent had caused the cholera had come from the outside in.

When Whitehead expressed his gratitude for the book, he offered Snow one "*a priori* objection" to the contamination theory: if a specific case of cholera had started the outbreak, then shouldn't the cholera's rapid diffusion through the surrounding population have made the water increasingly deadly over the course of the week, as more and more rice-water evacuations passed into the well water? If

Snow's theory was correct, Whitehead continued, the pattern of the outbreak should have been a gradual upward slope, rather than a sudden spike followed by a steady decline. And then there was the matter of the contamination route. The Paving Board had found no communication between the Broad Street well and the local sewers. The idea of a cesspool contaminating the well seemed even more ludicrous to Whitehead. As far as the curate knew, all the cesspools had been eliminated since the passing of the Nuisances Act.

But Snow's monograph and the growing stockpile of data had pushed Whitehead closer and closer to accepting the waterborne theory. If Snow was right, there had to be, in the language of modern epidemiology, an index case, an original cholera victim whose evacuations had somehow found their way into the Broad Street well. Assuming an incubation period of a few days—enough time for the *V. cholerae* to find its way into the well and then into the small intestines of the first wave of sufferers—patient zero should have fallen ill sometime around the twenty-eighth of August. Whitehead went back and studied the *Weekly Returns* for the weeks before the outbreak, and found only two cases in the neighborhood: one death on the twelfth, and one on the thirtieth. On further investigation, both cases turned out to have transpired too far from the Broad Street well to have had any likely connection to the water there.

For several weeks, Whitehead was at an impasse. All the evidence that he had compiled pointed to the existence of an index case that would confirm, once and for all, the very theory that he had so long resisted. He was now almost convinced that the well had been contaminated, and that the famously pure waters of the Broad Street pump had been responsible for the devastation in his parish. But who had done the contaminating?

When Whitehead wasn't performing his duties at St. Luke's or interviewing the dispersed former residents of Broad Street, he could often be found sifting through the files at the Registrar-General's Office. The broad-brush statistics of the *Weekly Returns* had long since lost their utility for Whitehead; he needed the additional specificity that the original records offered. During one visit, while searching for some other stray detail, a record from the Broad Street files caught his eye: "At 40, Broad Street, 2d September, a daughter, aged five months: exhaustion, after an attack of diarrhoea four days previous to death."

Whitehead was already familiar with the sad story of baby Lewis. Her death had long been included in his chronology of the outbreak. What caught his eye this time was the commentary at the end: ". . . after an attack of diarrhoea four days previous to death." It had never occurred to Whitehead that an infant could have survived for more than a day or two with a disease that had killed many a grown adult in a matter of hours. But if baby Lewis had been sick for four days, that meant her illness would have predated the outbreak by at least a day. He knew at a glance that the address—40 Broad—put baby Lewis as close to the pump as anyone in the neighborhood.

Whitehead immediately dropped his other inquiry and rushed back to Broad Street, where he found Mrs. Lewis at home and willing to entertain further questions from the curate. She told him that her daughter had in fact been attacked a day earlier than Farr's record suggested: on the twenty-eighth, *five* days before her eventual demise. When Whitehead asked how she disposed of the baby's soiled diapers, she said the cloths were steeped in pails of water, some of which were then tossed into a sink in the backyard. But some she had dumped into a cesspool that lay in the basement at the front of the house.

The Reverend Whitehead could feel the chain of events click

into place. The case of baby Lewis matched the profile of the index case perfectly: an attack of cholera that occurred three days before the first wave of the general outbreak, where the victim's evacuations were deposited a matter of feet from the Broad Street well. It was exactly as John Snow had predicted. Whitehead convened the Vestry Committee immediately, and the men reached an easy agreement. The Broad Street well would be examined once again.

A local surveyor by the name of York was assigned the task of overseeing the second excavation of the Broad Street well. But this time the cesspool at the base of 40 Broad would be examined, too. Number 40 had a waste pipe connecting to the sewer, but the design was flawed on a number of levels. The cesspool at the front of the house had been intended to function as a trap, but in practice it served as a dam that blocked the normal flow into the sewer. Whitehead would later say that York had found there "abominations, unmolested by water, which I forebear to recite." The walls of the cesspool were lined with bricks that were so decayed that they could be "lifted from their beds without using the least force." Two feet and eight inches from the outer edge of the brickwork lay the Broad Street well. At the time of the excavation, the water line in the well was eight feet below the cesspool. Between the cesspool and the well, York reported finding "swampy soil" saturated with human filth.

The original excavation had missed all this because, guided by Benjamin Hall's dictates, it had examined only the interior of the well, and focused much of its inquiry on the quality of the water. The miasmatists from the Board of Health weren't interested in transmission routes, in flows. They didn't see the outbreak as a relay network the way John Snow did. They were looking for a general property of uncleanliness in the neighborhood, not an index case. If the well had been partly responsible for the outbreak, then the flaw was surely to be found inside the well itself. It never occurred to the Board of Health that the well, though sound, could have "caught" the disease from another source. And so the Board's inspectors merely peered down the well and sampled the water. They never bothered to look beyond those decaying walls, never bothered to see the connections.

But York's excavation had unearthed the gruesome truth. The contents of a cesspool were seeping into the Broad Street well. Anything living in the intestinal tracts of the residents at 40 Broad had direct access to the intestines of about a thousand other human beings. That was all the *V. cholerae* had needed.

As the Vestry Committee put the final touches on its report, Whitehead stumbled across the explanation for his final objection to Snow's theory. If the Broad Street well had been contaminated by the neighborhood's waste, why didn't the well get even deadlier as more and more of the neighborhood came down with the cholera? Why didn't the epidemic follow an exponential growth pattern, with each new case making the contamination worse? York's excavation had offered half an explanation, by narrowing the focus to 40 Broad. Cholera victims living elsewhere in the neighborhood weren't emptying their pails into the Broad Street well, and so their illness had no effect on the quality of the water there. But five people had died at 40 Broad alone, including some of the very first cases: the tailor, Mr. G, and his wife. Why hadn't their evacuations drained back into the well water at the height of the epidemic, thus fanning the flames even higher?

The answer turned out to be a simple matter of architecture. Only the Lewis family had ready access to the cesspool at the front of the house. The other residents, living on the upper floors, tossed their waste out the windows into the squalid courtyard at the back of the house. There was, no doubt, a vast colony of *V. cholerae* lying in wait in the dark earth behind 40 Broad, passed on from the intestines of the newly dead. But no one ever tried to drink from the courtyard's foul soil, and so the chain of infection stopped there. The population of *V. cholerae* in Soho was exploding at unthinkable rates, but the connection between the bacteria and the Broad Street well had been cut off after baby Lewis died, because Mrs. Lewis had nothing left to deposit in the cesspool at the front of her house.

As Whitehead shared his discoveries with Snow over those early months in 1855, a quiet but profound friendship bloomed between the two men. Many years later, Whitehead recalled the "calm, prophetic" manner in which Snow described the future of their mutual investigation. "You and I may not live to see the day," Snow explained to the young curate, "and my name may be forgotten when it comes; but the time will arrive when great outbreaks of cholera will be things of the past; and it is the knowledge of the way in which the disease is propagated which will cause them to disappear."

WITH THE INDEX CASE IDENTIFIED, THE VESTRY COMMITTEE was now ready to issue its report, and it would be a thorough vindication of Snow's original hypothesis. They began by methodically debunking the other popular explanations that had circulated in the months since the outbreak: meteorological conditions, sewer air, the lingering blight of the pesthouse fields. The pestilence had not leveled a disproportionate blow against any specific industry, nor had it singled out an economic class: both upstairs and downstairs had been devastated. Sanitary houses had suffered as readily as unsanitary ones.

Only one explanation had withstood the committee's extensive probe:

> The Committee is unanimously of the opinion that the striking disproportionate mortality in the "cholera area" . . . was in some manner attributable to the use of the impure water of the well in Broad Street.

In embracing the waterborne theory, the committee went out of their way to take a pointed swipe at the miasma hypothesis. The sentences are formal Victorian, suitable prose for a serious committee report on a deadly event. But they are fighting words, nonetheless:

> The weight of both positive and negative evidence appears to be clearly and unmistakable in one direction viz.—to show that the water had some preponderating influence in determining an attack. . . . If it be urged, in explanation of an atmospheric influence, that Cholera might be conveyed exclusively to some by a partial distribution of an impure air, it may be replied that no consideration of the streets, local levels, sewergrates, house drains, or direction of the wind, will explain the existence of such partial atmospheric impurity, whereas the individual use of the water has been actually traced, and its consequences may not be unreasonable inferred.

The Vestry Committee's report on the Broad Street epidemic was, technically, the second institutional victory for Snow's waterborne theory, but it felt like the first. Snow had convinced the parish's Board of Governors to remove the pump handle, though they had hardly been persuaded by his argument. Yet his case against the pump had genuinely won over the Vestry Committee. Snow's theory had even withstood the assault of a committed debunker. The Reverend Whitehead had actively set out to disprove the theory, but he had been so thoroughly convinced by Snow's argument that he ended up supplying the evidence that ultimately closed the case. The prosecutor had turned out to be the defense's star witness.

SURELY HERE IS WHERE THE FOG OF MIASMA SHOULD LIFT, and science finally win out over superstition for good. But science rarely lands such decisive blows, and the Broad Street case was no exception. Within a few weeks of the Vestry Committee report, Benjamin Hall's team issued their account of the St. James cholera epidemic. Its verdict on Snow's theory was unequivocal—and unequivocally dismissive:

> In explanation of the remarkable intensity of this outbreak within very definite limits, it has been suggested by Dr. Snow, that the real cause of whatever was peculiar in the case lay in the general use of one particular well, situate at Broad Street in the middle of the district, and having (it was imagined) its waters contaminated with the rice-water evacuations of cholera patients.
>
> After careful inquiry, we see no reason to adopt this belief. We do not find it established that the water was contaminated in the manner alleged; nor is there before us any sufficient evidence to show, whether inhabitants of the district, drinking from that well, suffered in proportion more than other inhabitants of the district who drank from other sources.

We see no reason to adopt this belief. Of course the Board of Health Committee saw no reason. Their field of vision had been framed by the boundaries of miasma months before, when Benjamin Hall first outlined the committee's objectives. This blanket dismissal of Snow's theory seems like a colossal folly to us now, but these were not unreasonable men. They were not hacks, working surreptitiously for Victorian special-interest groups. They were not blinded by politics or personal ambition.

They were blinded, instead, by an idea.

> That such local uncleanliness prevailed most intensely throughout the suffering districts, is evident from the reported results of house-to-house visitation. The exterior atmosphere was offensive with effluvia from ill-conditioned sewers; the houses were almost universally affected in the same manner, partly from the same source, partly from their own extreme defects of drainage and cleanliness, partly from unregulated slaughtering and other offensive trades; the inhabitants were overcrowded, perhaps to the greatest degree known even in London, and the general architecture of the locality was such as to render it almost insusceptible of ventilation.
>
> On the principle to which we have referred, and which we believe to be commonly recognised as presenting the most probable theory of choleraic irruptions, it will be obvious that the locality, notwithstanding its high level, contained every predisposing condition which (given the exciting cause) should render it prone to a violent epidemic explosion; and we believe that any person conversant with the laws of disease might have predicted its extreme liability to suffer what afterwards befell it.

Here is the logic of the Cholera Commission's report, paraphrased in plain English: "Cholera thrives in unventilated, crowded spaces where unsanitary conditions and noxious smells abound. We have examined the Broad Street area, and found it to be an unventilated, crowded space where unsanitary conditions and noxious smells abound. What more do you need?"

If there weren't human lives at stake, the Cholera Commision's report would be almost comical reading, capturing in excruciating detail the Gradgrindian overanalysis of utterly meaningless data. The first hundred pages read like a weather almanac, with dozens of tables documenting every atmospheric variable known to science. The section headings read as follows:

Atmospheric Pressure
Temperature of the Air
Temperature of the Thames Water
Humidity of the Air
Direction of the Wind
Force of the Wind
Velocity of the Air
Electricity
Ozone
Rain
Clouds
Comparison of the Meteorology of London, Worcester, Liverpool, Dunino, and Arbroath
Wind
Ozone [again]
Progress of the Cholera in the Metropolitan Districts in the Year 1853
Atmospheric Phenomena in the Year 1853
Atmospheric Phenomena in relation to Cholera in the Metropolitan Districts in the Year 1854

This litany makes it clear why the committee found no reason to believe Dr. Snow's theory. They were not, strictly speaking, investigating Dr. Snow's theory. Perhaps if they had spent a little more time investigating the patterns of water consumption on Broad Street, and a little less time compiling data on the meteorology of Dunino, they might have found Snow's argument more compelling.

The only concession the committee made to Snow's theory was a brief reference to the case of Susannah Eley. It was impossible to avoid the conclusion that the Broad Street water had been the vehicle of contamination in that instance. But the *experimentum crucis* was apparently not *crucial* enough for the miasmatists on the committee:

> The water was undeniably impure with organic contamination; and we have already argued that if, at the times of epidemic invasion there was operating in the air some influence which converts putrefiable impurities into a specific poison, the water of the locality, in proportion as it contains such impurities, would probably be liable to similar poisonous conversion.

This is circular argumentation at its most devious. The committee begins with the assertion that cholera is transmitted via the atmosphere. When it discovers evidence that contradicts this initial assertion—a clear case that cholera has been transmitted by water—the counter-evidence is invoked as further proof of the original assertion: the atmosphere must be so poisoned that it has infected the water as well. Psychologists call this type of faulty reasoning "confirmation bias": the tendency to force new information to fit one's preconceptions about the world. For Benjamin Hall's committee, the confirmation bias toward miasma was so strong that it literally blinded them to the patterns that Snow and Whitehead perceived so clearly—blinded them on two fundamental levels. Hall's initial biases had structured the inquiry in such a way that most of the relevant data never came before the committee. And when a few telltale patterns did slip through the cracks, the committee was so conceptually mired in the prevailing model that it turned the waterborne theory's *experimentum crucis* into yet another testament to the power of miasma.

And so the miasma theory did not crumble immediately after the Broad Street outbreak, though its days were numbered. Eventually, Snow's and Whitehead's parallel investigations would be seen as the turning point in the battle against cholera. But it would take yet another outbreak—more than a decade later—for that narrative to take hold for good.

It is not known if Sarah Lewis ever learned that the final days she spent tending to her daughter had triggered the most devastating outbreak in the history of London. If so, the weight of the news must have been unbearable, because the outbreak she had unwittingly set in motion eventually killed her husband as well. Thomas Lewis had fallen ill that Friday, September 8, within hours of the pump handle's removal. He fought the disease much longer than most, surviving for eleven days. The young policeman finally succumbed on the nineteenth of September, leaving a childless widow alone in a ruined neighborhood. The outbreak had begun at 40 Broad Street, and it ended there as well.

The timing of Thomas Lewis' illness suggests one chilling alternative history. The Broad Street outbreak had subsided in part because the only viable route between the well and the neighborhood's small intestines had run through the cesspool at 40 Broad. When baby Lewis died, the connection had died with it. But when her husband fell ill, Sarah Lewis began emptying the buckets of soiled water in the cesspool all over again. If Snow had not persuaded the

Board of Governors to remove the handle when he did, the disease might have torn through the neighborhood all over again, the well water restocked with a fresh supply of *V. cholerae.* And so Snow's intervention did not just help bring the outbreak to a close. It also prevented a second attack.

MARLBOROUGH STREET
POLAND STREET
NOEL
PORTLAND STREET
BERWICK STREET
WARDOUR MEWS
WORK HOUSE
PORTLAND MEWS
BENTINCK STREET
EDWARD ST
MARSHALL ST
WEST STREET
MARSHALL STREET
DUFOURS PLACE
BROAD STREET
PUMP
BREWERY
NEW STREET
HOPKINS STREET
HUSBAND ST
WILLIAM AND MARY YARD
LITTLE WINDMILL STREET
GREAT PULTENEY STREET
BRIDLE STREET
SILVER STREET
UP. JAMES ST
UP. JOHN ST
GOLDEN SQUARE
WARWICK STREET
PUMP
LEICESTER ST
LOW. JAMES ST
PUMP
HAM YARD

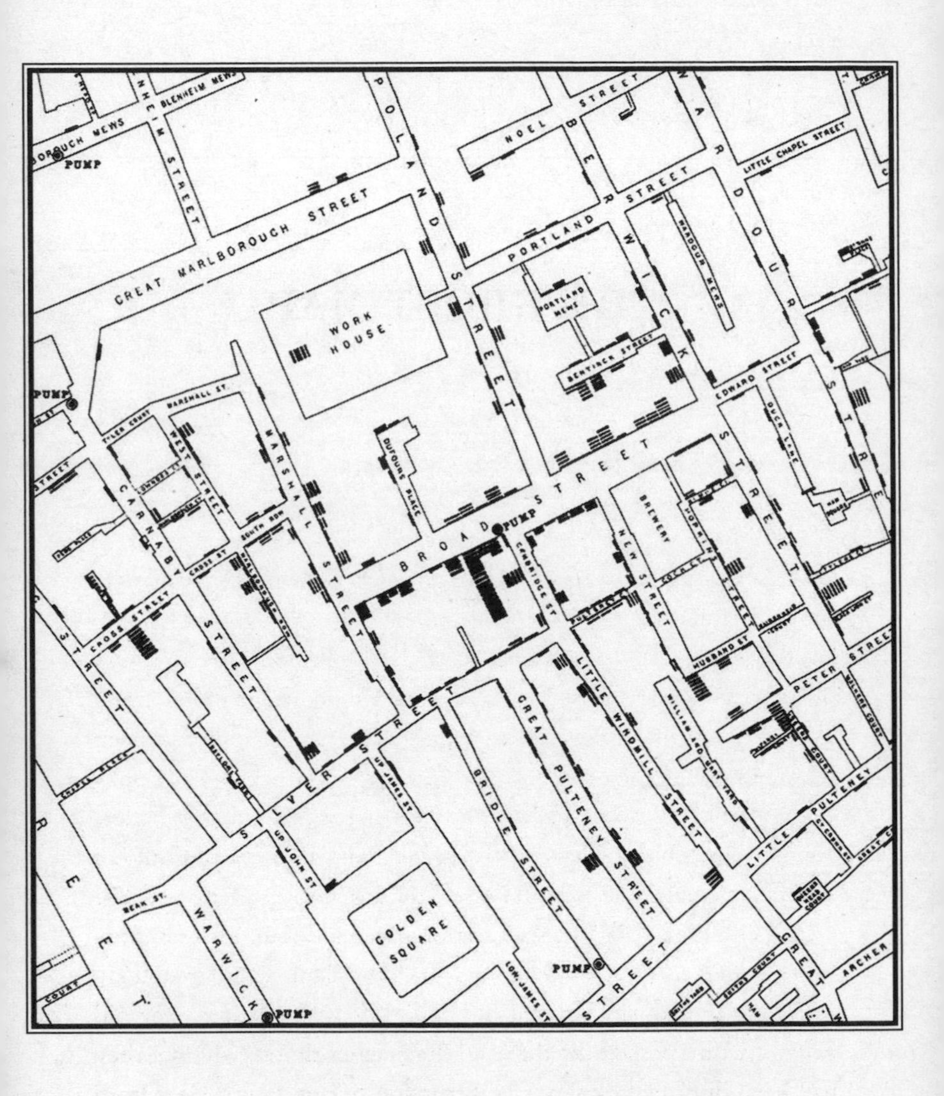
BLENHEIM MEWS
PUMP
GREAT MARLBOROUGH STREET
POLAND STREET
NOEL STREET
LITTLE CHAPEL STREET
PORTLAND STREET
PORTLAND MEWS
BENTINCK STREET
WORK HOUSE
MARSHALL ST.
TYLER COURT
WEST STREET
MARSHALL STREET
CARNABY STREET
DUFOURS PLACE
BROAD STREET
PUMP
CAMBRIDGE ST
BREWERY
NEW STREET
EDWARD STREET
DUCK LANE
HOPKINS STREET
SOUTH ROW
CROSS ST
CROSS STREET
SILVER STREET
GREAT PULTENEY STREET
LITTLE WINDMILL STREET
BRIDLE STREET
WILLIAM AND MARY YARD
HUSBAND ST
PETER STREET
WALKERS COURT
LITTLE PULTENEY
UP JAMES ST
UP JOHN ST
BEAK ST.
WARWICK
PUMP
GOLDEN SQUARE
LOW. JAMES ST
PUMP
STREET
GREAT
ARCHER

Conclusion

THE GHOST MAP

IN THE FIRST FEW DAYS AFTER THE PUMP HANDLE'S REMOVAL, an engineer by the name of Edmund Cooper began examining the Broad Street epidemic on behalf of the Metropolitan Commission of Sewers. Rumors that sewer excavations had unearthed the decaying but still pestilent corpses from the plague burial grounds had been buzzing through the neighborhood. Even the newspapers had implicated the old pesthouse fields. (The *Daily News* had published a letter on September 7 accusing the sewer builders of unearthing an "immense quantity of human bones" during their excavations in the area.) With these scandalous accusations floating about, the Commission dispatched Cooper to investigate the claim. Cooper quickly arrived at the conclusion that the bodies of two-hundred-year-old plague victims posed little threat to the neighborhood, whether they had been disturbed by sewer construction or not. It was clear from the *Weekly Returns*—and from Cooper's on-scene investigating—that the sewer construction had not likely played a role, given the geographic dispersal of deaths. But Cooper needed a way to represent these patterns in an intelligible manner that both the laypeople of the neighborhood and his supervisors might understand. So he created a map of the outbreak. He modified an existing plan of the neighborhood that showed the new sewer lines, adding visual codes to indicate both the location of cholera deaths and the site of the original plague pit. For each house that had suffered a loss, Cooper drew a black bar by the address, followed by a succession of thin lines indicating how many deaths had occurred at that address. In the northwest corner of the map, roughly centered over Little Marlborough Street, Cooper drew a circle inscribed with the words "Supposed Location of Plague Pit." A quick glance at the map made it clear that the outbreak had been triggered by some other source: the deaths were concentrated several blocks to the southeast of the ancient burial site. Only a handful of deaths had occurred within Cooper's circle, and the houses to the immediate south and east of the circle had been spared entirely. If some noxious effluvium had risen out of the plague pit, surely the residents living directly on top of the pit would have suffered the worst casualty rate.

Cooper's original layout would be copied and expanded in another map produced for the Board of Health investigation that included data from the more extensive survey that had been carried out that fall. Once again, the map exonerated the plague pit, though the committee ultimately included the sewer lines as a potential source of miasmatic poisoning in the area. Both maps were well-crafted specimens of the new art of dot mapping—that is, representing the spatial path of an epidemic by marking each case with dots (or bars) on a map. They were both attempts to tell the story of the Broad Street outbreak from the bird's-eye view, to see the patterns of the disease as it erupted through the neighborhood. They were both superbly detailed: old and new sewer lines were documented with distinct markings; each gulley hole was represented by an icon on the map, along with ventilators and side entrances and the street number of every house in the parish. Even the neighborhood pumps were included. But as exacting as Cooper's map was, it ultimately had too much detail to make sense of the story. The connection between the Broad Street pump and the surrounding deaths was lost under the sheer mass of data that Cooper had charted. For a map to explain the true cause behind the Broad Street outbreak, it needed to show less, not more.

—

JOHN SNOW BEGAN WORKING ON HIS FIRST MAP OF THE Broad Street outbreak sometime in the early fall of 1854. Its initial form, which he shared publicly at a meeting of the Epidemiological Society in December, resembled Cooper's survey, with two small modifications: Each death was represented by a thick black bar, which made the houses that had suffered significant deaths more vivid on the map. And the detail on the map was reduced, with everything but the basic street layout and the icons representing the thirteen public water pumps that served the greater Soho area eliminated. The visual impact of the map was striking. Because it represented a larger section of London—from Hanover Square in the west to Soho Square in the east, and all the way south to Piccadilly Circus—eleven of the pumps were shown to be entirely clear of local cholera cases. The Little Marlborough Street pump had a few black bars in its immediate vicinity, but they were nothing compared with the concentration of death around the Broad Street pump, black bars lining the nearby streets like solemn high-rises. Without a prominent icon for the

Broad Street pump, the other dot maps of the epidemic had presented an amorphous shape, like a cloud hanging over western Soho. But when you emphasized the pumps in the image, the map took on a new clarity. Cholera wasn't lingering over the neighborhood in a diffuse form. It was radiating out from a single point.

In effect, Snow had given the death and darkness of the Broad Street outbreak a new kind of clarity. His first map has been rightly celebrated for its persuasiveness, and variations of it have been reproduced in countless textbooks on cartography, information design, and public health. A landmark 1911 textbook on epidemiology, *Sedgwick's Principles of Sanitary Science and Public Health,* included a dozen pages on the Broad Street case and featured a revised version of the map prominently. Thanks to that continued attention, the map has become the defining symbol of the entire Broad Street outbreak. But its significance has been somewhat misunderstood. The black bars marking the ghosts of Soho were a striking visual element, but they were hardly Snow's invention. Not only had dot maps been created to visualize previous cholera outbreaks, but at least one (Cooper's) had already been created to document the Broad Street outbreak itself before Snow began work on his map. Part of what made Snow's map groundbreaking was the fact that it wedded state-of-the-art information design to a scientifically valid theory of cholera's transmission. It was not the mapmaking technique that mattered; it was the underlying science that the map revealed.

Snow modified his original map for publication in two places—the Vestry Committee's report and the second edition of his own monograph on cholera. Augmented by the new data on the outbreak that Whitehead and others had assembled, the second version of the map contained Snow's most significant contribution to the field of disease mapping. (Ironically, it goes unmentioned in Edward

Tufte's extensive account of Snow's mapmaking in *Visual Explanations,* which almost single-handedly elevated Snow's work to the information-design canon.) After presenting to the Epidemiological Society, Snow had realized that his original map was still vulnerable to a miasmatic interpretation. Perhaps the concentration of deaths around the Broad Street pump was merely evidence that the pump was releasing noxious fumes into the air. And so Snow realized he needed a way to represent graphically the foot-traffic activity around the pump that he had so painstakingly reconstructed. He needed to show *lives,* not just deaths; he needed to show the way the neighborhood was actually traversed by its residents.

To solve this problem, Snow drew upon a centuries-old mathematical tool that would later be termed the Voronoi diagram. (It is unlikely that Snow knew anything of the device's history, though he was certainly the first to apply it to disease mapping.) A Voronoi diagram conventionally takes the shape of a two-dimensional field made up of points surrounded by "cells." The cells define the region around each point that is closer to that particular point than any other point in the diagram. Imagine a football field with a point on each goal line. The Voronoi diagram of that field would be divided into two cells, the demarcation between them being at the fifty-yard line. If you stand anywhere on the field on the home-team side of the fifty-yard line, you are closer to the point on the home team's goal line than you are to the point on the other goal line. Most Voronoi diagrams, of course, involve many points scattered about in unexpected ways, resulting in a honeycomb pattern of cells surrounding their local points.

What Snow set out to do with his second map was to create a Voronoi diagram using the thirteen pumps as points. He would diagram a cell that showed the exact subsection of addresses on the map that were closer to the Broad Street pump than they were to any other pump. But these distances would have to be calculated on foot-traffic terms, not the abstract distances of Euclidean geometry. The cell was warped by the erratic arrangement of streets in Soho. Some addresses were closer to Broad Street as the crow flies, but if you actually paced the routes out by foot, winding your way through the crooked alleys and side streets of Soho, another pump turned out to be closer. It was, as the historian Tom Koch astutely notes, a map organized as much around time as around space: instead of measuring the exact distance between two points, it measured how long it took to walk from one point to another.

And so the second version of the map—the one that made it into both Snow's monograph and the Vestry report—included a slightly odd, wandering line that circumscribed the center of the outbreak, roughly in the shape of a square with five or six areas jutting out, like small peninsulas, into the surrounding neighborhood. This was the area encompassing all those residents for whom the quickest trip for water was to the Broad Street pump. Superimposed over the black bars that marked each death, the amorphous shape took on sudden clarity: each peninsula extended out to embrace another distinct cluster of deaths. Beyond the circumference of the cell, the black bars quickly disappeared. Snow's visual case for his waterborne theory revolved around a striking correspondence between two shapes: the shape of the outbreak area itself, and the shape of best proximity to the Broad Street pump. If the cholera were somehow spreading as a miasmatic emission from the pump, the shape of the neighborhood deaths would have looked quite different: not a perfect circle, perhaps, since some houses would have been more vulnerable than others. But it certainly wouldn't have followed so precisely the contours of street-level (i.e., foot-traffic) proximity to the Broad Street well.

The miasma wouldn't be influenced by the eccentricities of street layouts, after all, and it certainly wouldn't be influenced by the location of other neighborhood pumps.

And so the ghosts of the Broad Street outbreak were reassembled for one final portrait, reincarnated as black bars lining the streets of their devastated neighborhood. In dying, they had collectively made a pattern that itself pointed to a fundamental truth, though it took a trained hand to make that pattern visible. And yet, however elegant its design, the map's immediate influence was far less dramatic than folklore has it. The map didn't solve the mystery of the outbreak. It didn't lead to the pump handle's removal and thus bring an end to the epidemic. In fact, it failed to sell the Board of Health on the merits of the waterborne theory. Yet despite those reservations, Snow's map deserves its iconic status. The case for the map's importance rests on two primary branches: its originality and its influence.

The originality of the map did not revolve around the decision to map an epidemic, or even the decision to encode deaths in bars etched across the street diagram. If there was a formal innovation, it was that wobbly circumference that framed the outbreak in the second version, the Voronoi diagram. But the real innovation lay in the data that generated that diagram, and in the investigation that com piled the data in the first place. Snow's Broad Street map was a bird's-eye view, but it was drawn from true street-level knowledge. The data that it sketched out in graphic form was a direct reflection of the ordinary lives of the ordinary people who made up the neighborhood. Any engineer could have crafted a dot map from William Farr's *Weekly Returns.* But the Snow map drew on a deeper, more intimate, source: two Soho residents talking to their neighbors, walking the streets together, sharing information about their daily routines, and tracking down the long-departed émigrés. Neighborhood demographics had been projected onto maps before, of course, but invariably the projections involved the official interventions of the census takers or the Board of Health. Snow's map—with Whitehead's local knowledge animating it—was something else entirely: a neighborhood representing itself, turning its own patterns into a deeper truth by plotting them on a map. The map is a brilliant work of information design and epidemiology, no doubt. But it is also an emblem of a certain kind of community—the densely intertwined lives of a metropolitan neighborhood—an emblem that, paradoxically, was made possible by a savage attack on that community.

As for influence, it's pretty to think of John Snow unveiling the map before the Epidemiological Society to amazed and thunderous applause, and to glowing reviews in *The Lancet* the next week. But that's not how it happened. Its persuasiveness seems obvious to us now, living as we do outside the constraints of the miasma paradigm. But when it first began circulating in late 1854 and early 1855, its impact was far from dramatic. Snow himself seems to have thought that his South London Water Works study would ultimately be the centerpiece of his argument, the Broad Street map merely a piece of supporting evidence, a sideshow.

The tide of scientific opinion would eventually turn in Snow's favor, and when it turned, the Broad Street map grew in stature. Most accounts of the outbreak reproduced the map in some fashion—so often, in fact, that copies of copies began appearing in textbooks, described erroneously as original reproductions. (Most of them lacked the critical Voronoi diagrams.) As the waterborne theory of cholera became increasingly accepted, the map was regularly invoked as a shorthand explanation of the science behind the theory. It was easier to point to those black bars emanating ominously from the pump than it was to explain the whole idea of microorganisms invisible to the human eye. The map may not have had the impact on its immediate audience that Snow would have liked, but something about it reverberated in the culture. Like the cholera itself, it had a certain quality that made people inclined to reproduce it, and through that reproduction, the map spread the waterborne theory more broadly. In the long run, the map was a triumph of marketing as much as empirical science. It helped a good idea find a wide audience.

SNOW'S MAP MAY HAVE HAD A CRUCIAL SHORT-TERM IMPACT as well, though this is closer to an inference than an empirical fact. We know that Henry Whitehead's interest in the waterborne theory turned decisively after Snow gave him a copy of his revised cholera monograph in the late winter of 1855. That monograph contained the second edition of Snow's map. It's entirely possible that seeing all those deaths radiating out from the Broad Street pump played a role in changing the curate's mind. He had spent more time than anyone working through the intimate details of those lives and deaths—first attending the sick as a clergyman, then investigating the outbreak as an amateur detective. It must have been a revelation to see all that data rendered from above for the first time.

Persuading an assistant curate of the merits of the waterborne theory might seem like a minor accomplishment. But Whitehead's investigations in 1855 were ultimately as decisive as Snow's in solving the Broad Street mystery. His "conversion experience" reading Snow's monograph set him off in search of the index case, eventually leading him to baby Lewis. The discovery of baby Lewis led to York's excavation of the pump, which confirmed a direct connection between the pump and the cesspool at 40 Broad.

It's conjecture, of course, but it's nonetheless entirely reasonable to assume that without the Reverend Whitehead's contributions the Vestry Committee would have never blamed the outbreak on the Broad Street pump. Without an index case and an unequivocal link to the well water, without the support of one of the neighborhood's most beloved characters, it would have been so much easier for the Vestry Committee to equivocate, to blame the outbreak on the neighborhood's generally pitiful sanitary standards—in the streets and in the houses, in the water and in the air. It would have been so much easier for the Vestry Committee to fall back on the miasmatic haze of the Board of Health report. But the final compilation of evidence had been too overwhelming for such stock explanations. When you combined Snow's original data with Whitehead's more exhaustive investigation, when you factored in the index case and the decaying brickwork, the conclusion was inevitable: the pump was the source of the outbreak.

The Vestry Committee's verdict meant that for the first time an official committee investigation had endorsed the waterborne theory. It was a small victory, since the vestry had no power over public-health issues outside Soho, but it gave Snow and his future allies something that Snow had long sought: an official endorsement. In the years and decades that followed, the Vestry Committee report grew in influence as the story of the Broad Street outbreak was retold. Slowly, over time, it occluded the Board of Health investigation altogether. The twelve pages devoted to Broad Street in *Sedgwick's Principles of Sanitary Science and Public Health* quote extensively from the Vestry report, while the Board of Health verdict goes unmentioned. The vast majority of the retellings of the Broad Street case fail to mention the signal fact that among the public-health authorities of the day, Snow's investigation was of no significance.

Rewinding the tape of history and imagining alternative scenarios is always a fanciful exercise, but it can be instructive. If the Vestry Committee had not endorsed the waterborne theory, then the Broad Street episode would likely have entered the historical record as yet another example of miasma's deadly reach: a crowded, unsanitary neighborhood suffused with hideous smells that got its comeuppance. Snow's interventions would have remained the work of an illustrious maverick, an outsider with an unproven theory who failed to convince anyone other than a panicked Board of Governors that had removed a pump handle out of desperation. No doubt science would ultimately have come around to the waterborne theory, but it might well have taken decades longer without the clarity and reproducibility of the Broad Street story and its accompanying map. How many thousands more might have died in that interval?

It is a subtle chain of causal connections, but a plausible one nonetheless. The map helps tip Whitehead toward the waterborne theory, which prods him to discover the index case, which necessitates the second excavation, which ultimately tips the Vestry Committee toward Snow's original theory. And the endorsement of the Vestry Committee rescues Broad Street from the side of the miasmatists. It becomes the most powerful and seductive proxy for Snow's waterborne theory, thus accelerating the adoption of the theory by the very same public-health institutions that had renounced it so thoroughly at the time of the outbreak. The map may not have persuaded Benjamin Hall of the dangers of contaminated water in the spring of 1855. But that doesn't mean it didn't change the world in the long run.

Imagining the chain of events this way makes one fact overwhelmingly clear: John Snow may have been instrumental in first identifying the pump as the likely culprit behind the outbreak, but Whitehead ultimately supplied the crucial evidence for establishing the pump's role. The shorthand version of the Broad Street case invariably settles on the image of the visionary scientist, working alone against the dominant paradigm, discovering the secret cause behind a terrible plague. (Whitehead is often mentioned in popular accounts, but usually as a sort of dutiful apprentice, helping Snow with the door-to-door surveys.) But Broad Street should be understood not just as the triumph of rogue science, but also, and just as important, as the triumph of a certain mode of engaged amateurism. Snow himself was a kind of amateur. He had no institutional role where cholera was concerned; his interest in the disease was closer to a hobby than a true vocation. But Whitehead was an amateur par excellence. He had no medical training, no background in public health. His only credentials for solving the mystery behind London's most devastating outbreak of disease were his open and probing mind and his intimate knowledge of the community. His religious values had brought him into close contact with the working poor of Soho, but they had not blinded him to the enlightenments of science. If part of the significance of Snow's second map lay in the way it empowered the community to represent itself, Whitehead was the conduit that made that representation possible. Whitehead was not an expert, an official, an authority. He was a local. That was his great strength.

And here lies an antidote of sorts to the horror of Broad Street, to the grisly image of entire families dying together in their single-room flats: the image of Snow and Whitehead building an unlikely friendship in those late winter months of 1855, drawn together by a terrible outbreak of disease in their neighborhood and, ironically, by Whitehead's initial skepticism about Snow's theory. We know very little about the personal interaction between the two men, beyond the crucial data they exchanged, beyond Snow's sharing of his monograph, and his prophetic statements about the future of cholera. But it is clear from Whitehead's subsequent recollections that a powerful bond formed between them—the quiet, awkward anesthesiologist and the compulsively social curate—a bond forged both by living through an urban battleground of unimaginable terror, but also by jointly unearthing the secret cause behind the carnage.

This is not mere sentimentality. The triumph of twentieth-century metropolitan life is, in a real sense, the triumph of one image over the other: the dark ritual of deadly epidemics replaced by the convivial exchanges of strangers from different backgrounds sharing ideas on the sidewalk. When John Snow first stepped up to the Broad Street pump in early September 1854, it was by no means clear which image would be victorious. London seemed to be destroying itself. You could leave town for a weekend and come back to find ten percent of your neighbors being wheeled down the street in death carts. That was life in the big city.

Snow and Whitehead played a small but defining role in reversing that trend. They solved a local mystery that led, ultimately, to a series of global solutions—solutions that transformed metropolitan living into a sustainable practice and turned it away from the collective death drive that it threatened to become. And it was precisely their metropolitan connection that made this solution possible: two strangers of different backgrounds, joined by circumstance and proximity, sharing valuable information and expertise in the public space of the great city. The Broad Street case was certainly a triumph of epidemiology, and scientific reasoning, and information design. But it was also a triumph of urbanism.

John Snow would never get to experience that triumph in its entirety. In the first few years that followed the outbreak, supporters of the waterborne theory grew in number and in visibility. Snow's monograph had included both the Broad Street case and the South London water-supply study, and the combination seemed to produce converts at a much greater clip than the original monograph had six years before. John Sutherland, prominent inspector for the Board of Health, made several public statements that offered at least a partial endorsement of the waterborne theory. William Farr's *Weekly Returns* grew increasingly supportive of the theory. Several publications appeared that argued for the waterborne theory without crediting Snow for the original insight—including a few that credited William Budd with the discovery of cholera's waterborne nature. Perhaps aware that his legacy would ultimately revolve around his cholera investigations, Snow responded to these papers with politic, but firm, letters to the medical journals, reminding his colleagues of his precedence in these matters.

Still, miasma retained its hold over many, and Snow himself was often subjected to derisive treatment by the scientific establishment. In 1855, he gave his testimony in Parliament on behalf of the "offensive trades" before a committee on the Nuisances Removal Act. Snow argued eloquently that infectious diseases were not spread through the foul smells emitted by the bone-boilers and gut spinners and tanners of industrial London. Again, he drew upon reasoned statistical analysis, arguing that the laborers who worked in these establishments would have had a much greater incidence of disease than the general public if the miasma were somehow breeding epidemics. The fact that they did not show a disproportionate rate of contagion—despite their immersion in the fumes—meant that the cause of disease lay elsewhere.

Benjamin Hall, ever the miasmatist, expressed open disbelief at Snow's testimony. Edwin Chadwick would shortly after denounce

Snow's reasoning as illogical. But the real assault would come in an unsigned editorial in *The Lancet* that tore into Snow with remarkable fury and disdain:

> Why is it, then, that Dr. Snow is so singular in his opinion? Has he any facts to show in proof? No! . . . But Dr. Snow claims to have discovered that the law of propagation of cholera is the drinking of the sewage-water. His theory, of course, displaces all other theories. Other theories attribute great efficacy in the spread of cholera to bad drainage and atmospheric impurities. *Therefore,* says Dr. Snow, gases from animal and vegetable decomposition are innocuous! If this logic does not satisfy reason, it satisfies a theory; and we all know that theory is often more despotic than reason. The fact is, that the well whence Dr. Snow draws all sanitary truth is the main sewer. His *specus,* or den, is a drain. In riding his hobby very hard, he has fallen down through a gully-hole and has never since been able to get out again.

The confidence of the miasmatists couldn't last forever. In June 1858, a relentless early-summer heat wave produced a stench of epic proportions along the banks of the polluted Thames. The press quickly dubbed it the "Great Stink": "Whoso once inhales the stink can never forget it," the *City Press* observed, "and can count himself lucky if he live to remember it." Its overwhelming odors shut down Parliament. As the *Times* reported on June 18:

> What a pity . . . that the thermometer fell ten degrees yesterday. Parliament was all but compelled to legislate upon the great London nuisance by the force of sheer stench. The intense heat had driven

our legislators from those portions of their buildings which overlook the river. A few members, bent upon investigating the matter to its very depth, ventured into the library, but they were instantaneously driven to retreat, each man with a handkerchief to his nose.

But a funny thing happened when William Farr calculated his weekly returns for those early weeks of June. The rates of death from epidemic disease proved to be entirely normal. Somehow the most notorious cloud of miasmatic air in the history of London had failed to produce even the slightest uptick in disease mortality. If all smell was disease, as Edwin Chadwick had so boldly pronounced more than a decade before, then the Great Stink should have conjured up an outbreak on the scale of 1848 or 1854. Yet nothing out of the ordinary had happened.

It's easy to imagine John Snow taking great delight in the puzzling data from the *Weekly Returns,* perhaps writing up a brief survey for *The Lancet* or the *London Medical Gazette.* But he never got the opportunity. He had suffered a stroke in his office on June 10, while revising his monograph on chloroform, and died six days later, just as the Great Stink was reaching its peak above the foul waters of the Thames. He was forty-five years old. More than a few friends wondered if his many experiments inhaling experimental anesethetics in his home lab had brought on his sudden demise.

Ten days later, *The Lancet* ran this brief, understated item in its obituary section:

DR. JOHN SNOW—This well-known physician died at noon on the 16th instant, at his house in Sackville-street, from an attack of apoplexy. His researches on chloroform and other anaesthetics were appreciated by the profession.

Snow might have hoped that cholera would prove central to his legacy, but in the first obituary that ran after his death it didn't even warrant a mention.

AFTER YEARS OF BUREAUCRATIC WAFFLING, THE GREAT STINK finally motivated the authorities to deal with the crucial issue that John Snow had identified a decade before: the contamination of the Thames water from sewer lines emptying directly into the river. The plans had been in the works for years, but the public outcry over the Great Stink had tipped the balance. With the help of the visionary engineer Joseph Bazalgette, the city embarked on one of the most ambitious engineering projects of the nineteenth century: a system of sewer lines that would carry both waste and surface water to the east, away from Central London. The construction of the new sewers was every bit as epic and enduring as the building of the Brooklyn Bridge or the Eiffel Tower. Its grandeur lies belowground, out of sight, and so it is not invoked as regularly as other, more iconic, achievements of the age. But Bazalgette's sewers were a turning point nonetheless: they demonstrated that a city could respond to a profound citywide environmental and health crisis with a massive public-works project that genuinely solved the problem it set out to address. If Snow and Whitehead's Broad Street investigation showed that urban intelligence could come to understand a massive health crisis, Bazalgette's sewers proved that you could actually do something about it.

North of the Thames, the plan for the new sewers involved three main lines, each at different levels of elevation, running eastward parallel to the river. On the south side, there were to be two main lines. All the city's existing surface water and waste lines would empty into one of these "intercepting" sewers, and the contents would then flow—and in some cases be pumped—several miles east of the city. On the north side, they drained into the Thames at Barking; on the south, the outfalls were located at Crossness. The sewers only discharged into the Thames during high tide, after which the seaward pull of low tide would flush the city's waste out to the ocean.

It was a demonically complicated undertaking, given that the city already had a complicated infrastructure of pipes and rail stations and buildings—not to mention a population of nearly three million people—that Bazalgette somehow had to work around. "It was certainly a very troublesome job," he would later write, with typical English understatement. "We would sometimes spend weeks in drawing out plans and then suddenly come across some railway or canal that upset everything, and we had to begin all over again." Yet somehow, the most advanced and elaborate sewage system in the entire world was largely operational by 1865. The numbers behind the project were staggering. In those six years, Bazalgette and his team had constructed eighty-two miles of sewers, using over 300 million bricks and nearly a million cubic yards of concrete. The main intercepting sewers had cost only £4 million to construct, which would be roughly $250 million today. (Of course, Bazalgette's labor costs were much cheaper than today's.) It remains the backbone of London's waste-management system to this day. Tourists may marvel at Big Ben or the London Tower, but beneath their feet lies the most impressive engineering wonder of all.

The best way to appreciate the scale of Bazalgette's achievement in person is to stroll along the Victoria or Chelsea embankments on the north side of the river, or along the Albert Embankment on the southern shore. Those broad, attractive esplanades were built to house the massive low-elevation interception lines that ran parallel to the Thames. Beneath the feet of those happy riverside pedestrians enjoying the view and the open air, beneath the cars hurtling along north of the river, there lies a crucial, hidden boundary, the last line of defense that keeps the city's waste from reaching the city's water supply.

That low-elevation northern sewer was one of the final lines to be completed, and the delays in building it turned out to play a determining role in London's last great outbreak of cholera. In late June 1866, a husband and wife living in Bromley-by-Bow in East London fell ill with cholera and died a few days later. Within a week a terrible outbreak of cholera erupted in the East End—the worst the city had seen since the ravages of 1853–1854. By the end of August, more than four thousand people had died. This time it was William Farr who did the first round of detective work. Puzzled by the sudden explosion of cholera in the city after a decade of relative dormancy, Farr thought of his old sparring partner, John Snow, and his surveys of the South London water companies that had brought Snow so regularly to the Registrar-General's Office. Farr decided to break down these new deaths along water-supply lines, and when he did, the pattern was unmistakable. The great majority of the dead had been customers of the East London Water Company. This time around, Farr wouldn't waste time with miasmatic objections. He didn't know *how* the East London supply had been contaminated, but clearly there was something deadly in that water. To waste time would be to condemn untold thousands to their deaths. Farr immediately ordered that notices be posted in the area advising residents not to drink "any water which has not been previously boiled."

Still, mysteries remained. Bazalgette's sewers were supposed to have cut off the fatal feedback loop between London's outputs and inputs, its waste and its water supply. And the East London Water

Company claimed to use extensive filtering at all of its reservoirs. If some contaminant had somehow made its way out of the city's sewers, it should have been picked up by the East London filters before being passed on to the wider population. Farr sent a letter to Bazalgette, who immediately wrote back, apologetically, to say that the drainage system in that part of the city had not been activated yet. "It is unfortunately just the locality where our main drainage works are not complete," he explained. The low-level sewer had been constructed, but Bazalgette's contractors had yet to finish the pumping station required to elevate the sewage so that gravity could continue to pull it down toward its ultimate outfall at Barking. And so the intercepting line in that area was not in use yet.

Attention then turned to the East London Water Company. Initially, company representatives swore that all their water had been run through state-of-the-art filter beds at their new covered reservoirs. But reports had surfaced of some customers discovering live eels in their drinking water, which suggested that the filters were not perhaps working optimally. An epidemiologist named John Netten Radcliffe had been assigned to investigate the outbreak, and he began looking into the filtering system in place at East London. Only a few months before, Radcliffe had read a memoir of the Broad Street outbreak authored by a curate who had played some role in the investigation. In the absence of John Snow, it occurred to Radcliffe that this individual might bring some valuable experience to this latest epidemic. And so the amateur epidemiologist Henry Whitehead was brought back to help solve one last case of poisoned water.

Radcliffe and Whitehead, along with other investigators, quickly uncovered a number of negligent practices at the East London company that had allowed the nearby River Lea to contaminate the groundwater around the company's reservoir at Old Ford. Eventually, the index cases at Bromley-by-Bow were tracked down; the doomed couple's water closet turned out to empty into the River Lea less than a mile from the Old Ford reservoir. In the end, the link to the East London water supply proved to be even more statistically pronounced than the link to the Broad Street pump had been in 1854. Ninety-three percent of the dead were eventually found to be East London Water customers.

This time, the verdict was nearly unanimous, and Snow's visionary research was widely acknowledged. Farr himself delivered some of the most powerful words in testimony before Parliament the year after the outbreak. He began in a satiric mode, deriding the commercial interests that sustained the miasma theory despite so much evidence to the contrary:

> As the air of London is not supplied like water to its inhabitants by companies, the air has had the worst of it both before Parliamentary Committees and Royal Commissions. For air no scientific witnesses have been retained, no learned counsel has pleaded; so the atmosphere has been freely charged with the propagation and the illicit diffusion of plagues of all kinds; while Father Thames, deservedly reverenced through the ages, and the water gods of London, have been loudly proclaimed immaculate and innocent.

Of course, one man *had* in fact served as "learned counsel" for the atmosphere, in much reviled testimony ten years before. And in turn, Farr acknowledged John Snow's defining role:

> Dr. Snow's theory turned the current in the direction of water, and tended to divert attention from the atmospheric doctrine. . . . The theory of the East wind with cholera on its wings, assailing the East

> End of London, is not at all borne out by experience of previous epidemics. . . . An indifferent person would have breathed the air without any apprehension; but only a very robust scientific witness would have dared to drink a glass of the waters of the Lea at Old Ford after filtration.

Farr's conversion to Snow's doctrine was so complete that he literally rewrote history to make it appear as though Snow's ideas had more initial success than they had actually enjoyed. In the introduction to his report on the 1866 outbreak, Farr, alluding to the investigation into the Broad Street case, delivers this stunning account of the Board of Health committee's findings:

> The final report of the scientific committee proved conclusively the extensive influence of water as a medium for the diffusion of the disease in its fatal forms. . . . Dr. Snow's view that the cholera-stuff was distributed in all its activity through water was confirmed. The special report . . . inculpated the Broad-street pump to some extent in the terrible outbreak of the St. James district. But the subject was further and more conclusively investigated by a committee, aided by Dr. Snow and by the Rev. Henry Whitehead.

Either Farr was willfully distorting the record, or—like so many subsequent accounts—his memory of the Vestry Committee's investigation had suppressed the Board of Health report. Recall the exact wording of the Board of Health's "confirmation" of Snow's theory: *"After careful inquiry, we see no reason to adopt this belief. We do not find it established that the water was contaminated in the manner alleged."* With confirmations like that, who needs criticisms?

Still, the waterborne hypothesis had at long last entered the dominant scientific paradigm. It pleased Whitehead to know that he had once again helped his old friend's ideas find a larger audience. Even *The Lancet* came around, editorializing in the weeks after the 1866 outbreak:

> The researches of Dr. Snow are among the most fruitful in modern medicine. He traced the history of cholera. We owe to him chiefly the severe induction by which the influence of the poisoning of water-supplies was proved. No greater service could be rendered to humanity than this; it has enabled us to meet and combat the disease, where alone it is to be vanquished, in its sources or channels of propagation. . . . Dr. Snow was a great public benefactor, and the benefits which he conferred must be fresh in the minds of all.

Apparently Dr. Snow found a way out of that "gully-hole" after all.

BY THE LAST DECADES OF THE NINETEENTH CENTURY, THE germ theory of disease was everywhere ascendant, and the miasmatists had been replaced by a new generation of microbe hunters charting the invisible realm of bacterial and viral life. Shortly after discovering the tuberculosis bacillus, the German scientist Robert Koch isolated *Vibrio cholerae* while working in Egypt in 1883. Koch had inadvertently replicated Pacini's discovery of thirty years earlier, but the Italian's work had been ignored by the scientific establishment, and so it was Koch who won the initial round of acclaim for identifying the agent that had caused so much trauma over the preceding century. History would come around to the Italian, though. In 1965, *Vibrio cholerae* was formally renamed *Vibrio cholerae Pacini 1854*.

Even these advances were not enough to convince a few remaining stalwarts—like Edwin Chadwick, who went to his grave in 1890 an unrepentant believer in the disease-causing powers of miasma. But most public-health institutions reoriented themselves around the new science. Establishing sanitary water supplies and waste-removal systems became the central infrastructure project of every industrialized city on the planet. The appearance of the electrical grid, around the turn of the century, tends to attract more attention, but it was the building of the invisible grid of sewer lines and fresh-water pipes that made the modern city safe for the endless consumer delights that electricity would bring. Bazalgette's project was a model for the world to emulate. By 1868, the pumping station at Abbey Mills was finally completed, which meant the northern branch of Bazalgette's grand sewer system was fully operational. By the mid-1870s, the entire system was online. Sewage continued to be pumped into the eastern end of the Thames until 1887, when the city began dumping waste into the open sea.

The changes ushered in by the sewer system were manifold: fish returned to the Thames; the stench abated; the drinking water became markedly more appetizing. But one change stood out above all the others. In all the years that have passed since Henry Whitehead helped track down the Old Ford reservoir contamination in 1866, London has not experienced a single outbreak of cholera. The battle between metropolis and microbe was over, and the metropolis had won.

Cholera would continue to terrorize Western cities into the first decades of the twentieth century, but with London's successful engineering project as a model, the outbreaks usually prodded the local authorities into modernizing their civic infrastructure. One such outbreak hit Chicago in 1885, after a heavy storm flushed the sewage collecting in the Chicago River far enough into Lake Michigan that it reached the intake system for the city's drinking water. Ten percent of the city's population died in the ensuing outbreak of cholera and typhoid, and the deaths ultimately led to the city's epic effort to reverse the flow of the Chicago River, sending the sewage away from the water supply. Hamburg had built a modern sewage system in the 1870s, modeled largely on London's, but the design had been flawed, and in 1892 cholera returned to claim nearly ten thousand lives out of a population one-seventh that of London. Because the major cholera epidemics of the preceding sixty years had all jumped the English Channel from Hamburg, Londoners waited anxiously as news of the German outbreak came over the wires. But their concern was unwarranted. Bazalgette's defenses held, and the cholera never appeared on British shores.

By the 1930s, cholera had been reduced to an anomaly in the world's industrialized cities. The great killer of the nineteenth-century metropolis had been tamed by a combination of science, medicine, and engineering. In the developing world, however, the disease continues to be a serious threat. A strain of *V. cholerae* known as "El Tor" killed thousands in India and Bangladesh in the 1960s and 1970s. An outbreak in South America in the early 1990s infected more than a million people, killing at least ten thousand. In the summer of 2003, damage to the water-supply system from the Iraq War triggered an outbreak of cholera in Basra.

There is a fearful symmetry to these trends. In many ways, the struggles of the developing world mirror the issues that confronted London in 1854. The megacities of the developing world are wrestling with the same problems of uncharted and potentially unsustainable growth that London faced 150 years ago. In 2015, the five largest cities on the planet will be Tokyo, Mumbai, Dhaka, São Paulo, and

Delhi—all of them with populations above 20 million. The great preponderance of that growth will be driven by so-called squatter or shantytown developments—entire sprawling cities developed on illegally occupied land, without any traditional infrastructure or civic planning supporting their growth. The scavenger classes of Victorian London have been reborn in the developing world, and their numbers are truly staggering. There are a billion squatters on earth now, and some estimates suggest that their numbers will double in the next twenty years. It's entirely possible that a quarter of humanity will be squatters by 2030. All the characters of the Victorian underground economy—the mud-larks and toshers and costermongers—may have largely disappeared from cities in the developed world, but everywhere else on the planet their numbers are exploding.

Squatter cities lack most of the infrastructure and creature comforts of developed metropolitan life, but they are nonetheless spaces of dynamic economic innovation and creativity. Some of the oldest shantytown developments—the Rocinha area in Rio de Janeiro, Squatter Colony in Mumbai—have already matured into fully functioning urban areas with most of the comforts we've come to expect in the developed world: improvised wood shacks giving way to steel and concrete; electricity; running water; even cable television. The main road in the squatter village of Sultaneyli in Istanbul is lined with six-story buildings, bustling with the commerce of ordinary city life: banks, restaurants, shops. And all of this has been accomplished without title deeds, without urban planners, without government-created civic infrastructure, on land that it is, technically speaking, illegally occupied. The squatter communities are not, by any measure, sinkholes of poverty and crime. They are where the developing world goes to get *out* of poverty. The writer Robert Neuwirth puts it best in his mesmerizing account of squatter culture, *Shadow Cities*: "With makeshift materials, they are building a future in a society that has always viewed them as people without a future. In this very concrete way, they are asserting their own being."

But that hope needs to be tempered with caution. The squatters still face significant obstacles. Arguably the most pressing obstacle is the one that confronted London a century and a half ago: the lack of clean water. Over 1.1 billion people lack access to safe drinking water; nearly 3 billion—almost half the planet—do not possess basic sanitation services: toilets, sewers. Each year 2 million children die from diseases—including cholera—that result directly from these unsanitary conditions. And so the megacities of the twenty-first century will have to learn all over again the lessons that London muddled through in the nineteenth. They'll be dealing with 20 million people, instead of 2 million, but the scientific and technological wisdom available to them far exceeds what Farr and Chadwick and Bazalgette had at their disposal.

Some of the most ingenious solutions now being proposed take us back to the waste-recycling visions that captivated so many Victorian minds. The inventor Dean Kamen has developed two affiliated machines—each the size of a dishwasher—that together can provide electricity and clean water to rural villages or shantytown communities that lack both. The power generator runs off a readily available fuel—cow dung—though Kamen says it will run off "anything that burns." Its output can power up to seventy energy-efficient bulbs. The ambient heat from the generator can be used to run the water purifier, which Kamen nicknamed Slingshot. The device accepts any form of water, including raw sewage, and extracts the clean water through vaporization. Kamen's prototype includes a "manual" featuring a single instruction: just add water. Just as the pure-finders once roamed London, recycling dog excrement for the leather tanners, the squatters of tomorrow may end up solving the sanitation problems of their community by using the very substances—animal and human waste—that cause the problems in the first place.

One cannot be unduly optimistic about how these megacities will face their potential crises in the coming years. There may be new technologies that enable the squatter communities to concoct public health solutions on their own, but governments will obviously need to play a role as well. It took industrial London a hundred years to mature into a city with clean water and reliable sanitation. The scavenger classes that Mayhew analyzed with such detail no longer exist in London, but even the wealthiest cities in the developed world continue to face problems of homelessness and poverty, particularly in the United States. But the developed cities no longer appear to be on a collision course with themselves, the way London did in the nineteenth century. And so it may take the megacities of the developing world a century to reach that same sense of equilibrium, and during that period there will no doubt be episodes of large-scale human tragedy, including cholera outbreaks that will claim far more lives than were lost in Snow's time. But the long-term prospects for urban life, even in these vast new sprawling "organisms," are solid ones. It's likely the megacities will mature faster than London did, precisely because of all the forms of expertise that were in embryo during the Broad Street events: epidemiology, public infrastructure engineering, waste management and recycling. And of course all that expertise has been greatly amplified by the connective powers of the Web, linking institutional knowledge with the local knowledge of amateurs to an extent that Snow and Whitehead could never have imagined.

It has never been easier for that local knowledge to find its way onto a map, making patterns of health and sickness (as well as less perilous matters) visible to experts and laypeople in new ways. The descendants of Snow's Broad Street map are now ubiquitous on the World Wide Web. Instead of Snow and Whitehead knocking on doors, and William Farr tabulating physicians' reports, we now have far-flung networks of health providers and government officials reporting outbreaks to centralized databases, where they are automatically mapped and published online. A service called GeoSentinel tracks infectious diseases among travelers; the CDC publish a weekly update on the current state of influenza in the United States, along with a near-endless array of charts and maps documenting the different strains of flu circulating through the national bloodstream. The popular ProMED-mail e-mail list offers a daily update on all the known disease outbreaks flaring up around the world, which surely makes it the most terrifying news source known to man. The technology has advanced dramatically, but the underlying philosophy remains the same: that there is something profoundly enlightening about seeing these patterns of life and death laid out in cartographic form. The bird's-eye view remains as essential as it was back in 1854. When the next great epidemic does come, maps will be as crucial as vaccines in our fight against the disease. But again, the scale of the observation will have broadened considerably: from a neighborhood to an entire planet.

The influence of the Broad Street maps extends beyond the realm of disease. The Web is teeming with new forms of amateur cartography, thanks to services like Google Earth and Yahoo! Maps. Where Snow inscribed the location of pumps and cholera fatalities over the street grid, today's mapmakers record a different kind of data: good public schools, Chinese takeout places, playgrounds, gay-friendly bars, open houses. All the local knowledge that so often remains trapped in the minds of neighborhood residents can now be translated into map form and shared with the rest of the world. As in 1854, the amateurs are producing the most interesting work, precisely because they have the most textured, granular experience of their community. Anyone can create a map that shows you where streets intersect and where hotels are; we've had maps like that for centuries. The maps now appearing are of a different breed altogether: maps of local knowledge created by actual locals. They're street-smart. They map the intangibles: blocks that aren't safe after dark, playgrounds that could use a renovation, local restaurants that have room for strollers, overvalued real estate offerings.

Even ordinary Web pages can be explored geographically now. Both Yahoo! and Google have established a standard convention for "tagging" a given piece of information—a blog post, say, or a promotional website—with geographic coordinates that are automatically interpreted by search engines. Someone writes into an online community forum with a complaint about a local park and tags the message with the park's exact location; someone writes up a minireview of a new restaurant; someone posts a notice about a summer sublet that they're offering. Up to now, all of those individual pieces of data possessed a location in the information space of the Web, in that they were associated with a URL—a "uniform resource locator." Now those items can possess a location in real-world space as well. In the near future, we'll use these geo-tags as we explore a new city, in much the way that we use search engines to explore the space of the Web today. Instead of looking for Web pages associated with a keyword or phrase, we'll look for pages associated with the streetcorner we're standing on. We'll be able to build instantly the kind of bird's-eye view of a neighborhood that Snow and Whitehead stitched together by hand over months of investigation.

These are technologies that thrive in urban centers, because they grow more valuable the more densely populated the environment is. A suburban cul-de-sac is unlikely to have a significant number of Web pages associated with it. But a streetcorner in a big city might well have a hundred interesting links: personal stories, reviews about the hot new bar around the corner, a potential date who lives three blocks away, a hidden gem of a bookstore—perhaps even a warning about a contaminated water fountain. These digital maps are tools for making new kinds of sidewalk connections, which is why they are likely to be less useful in communities without sidewalk culture. The bigger the city, the more likely it is that you'll be able to make an interesting link, because the overall supply of social groups and watering holes and local knowledge is so vast.

Jane Jacobs observed many years ago that one of the paradoxical effects of metropolitan life is that huge cities create environments where small niches can flourish. A store selling nothing but buttons most likely won't be able to find a market in a town of 50,000 people, but in New York City, there's an entire button-store district. Subcultures thrive in big cities for this reason as well: if you have idiosyncratic tastes, you're much more likely to find someone who shares those tastes in a city of 9 million. As Jane Jacobs wrote:

> Towns and suburbs . . . are natural homes for huge supermarkets and for little else in the way of groceries, for standard movie houses or drive-ins and for little else in the way of theater. There are simply not enough people to support further variety, although there may be people (too few of them) who would draw upon it were it there. Cities, however, are the natural homes of supermarkets and standard movie houses plus delicatessens, Viennese bakeries, foreign groceries, art movies, and so on, all of which can be found co-existing, the standard with the strange, the large with the small. Wherever

> lively and popular parts of cities are found, the small much outnumber the large.

The irony, of course, is that digital networks were supposed to make cities less attractive, not more. The power of telecommuting and instant connectivity was going to make the idea of densely packed urban cores as obsolete as the walled castle-cities of the Middle Ages. Why would people crowd themselves into harsh, overpopulated environments when they could just as easily work from their homestead on the range? But as it turns out, many people actually like the density of urban environments, precisely because they offer the diversity of Viennese bakeries and art movies. As technology increases our ability to find these niche interests, that kind of density is only going to become increasingly attractive. These amateur maps offer a kind of antidote to the scale and complexity and intimidation of the big city. They make everyone feel like a native, precisely because they draw upon the collective wisdom of the real natives.

City governments are exploring these new mapping technologies as well. Several years ago New York City rolled out its pioneering 311 service, which may well be the most radical enhancement of urban information management since William Farr's *Weekly Returns.* Modeled after the on-demand tech-support lines that New York mayor Michael Bloomberg built into the computer terminals that made him rich, as well as on a few smaller programs in cities such as Baltimore, 311 is ultimately three distinct services wrapped into one. First, it is a kinder, gentler version of 911; in other words, 311 is the number New Yorkers call when there's a homeless person sleeping near the playground—and not the number they call when someone's breaking into their apartment. (During the first year of 311's operation, the total number of 911 calls decreased for the first time in the city's history.) The 311 service also functions as a kind of information concierge for the city, offering on-demand information about all the city's services. Citizens can call to find out if the concert in Central Park has been canceled due to rain, if alternate-side parking is in effect, or the location of the nearest methadone clinic.

But the radical idea behind the service is that the information transfer is genuinely two-way. The government learns as much about the city as the 311 callers do. You can think of 311 as a kind of massively distributed extension of the city's perceptual systems, harnessing millions of ordinary "eyes on the street" to detect emerging problems or report unmet needs. (Bloomberg himself is notorious for calling in to report potholes.) During the blackout of 2003, many diabetic New Yorkers grew increasingly apprehensive about the shelf life of room-temperature insulin. (Insulin is traditionally kept refrigerated.) The city's emergency planners hadn't anticipated those worries, but within a matter of hours, Bloomberg was addressing the topic in one of the many press conferences broadcast over the radio that night. (Insulin, it turns out, remains stable for weeks at room temperature.) The insulin issue had trickled up the city's command chain thanks to calls into the 311 line. Those diabetics dialing in during the blackout got an answer to their question, but the city got something valuable in return: the calls had made them aware of a potential health issue that hadn't occurred to them before the lights went out.

Already 311 is having an impact on the city government's priorities. In the first year of operation, noise issues dominated the list of complaints: construction sites, late-night parties, bars and clubs spilling out onto sidewalks. The Bloomberg administration has subsequently launched a majority quality-of-life initiative combating city noise. Much as the COMPSTAT system revolutionized the way the police department fought crime by mapping problem areas with new precision, the 311 service automatically records the location of each incoming complaint in a massive Siebel Systems call-center database that feeds information throughout the city government. Geo-mapping software displays which streets have chronic pothole troubles and which blocks are battling graffiti.

Increase the knowledge that the government has of its constituents' problems, and increase the constituents' knowledge of the solutions offered for those problems, and you have a recipe for civic health that goes far beyond the superficial appeal of "quality of life" campaigns. When people talk about network technology revolutionizing politics, it's usually in the context of national campaigns: Internet fund-raising, or political blogging. But the most profound impact may be closer to home: keeping a neighborhood safe and clean and quiet, connecting city dwellers to the immense array of programs offered by their government, creating a sense that individuals can contribute to their community's overall health, just by dialing three numbers on a phone.

All of these extraordinary new tools are descendants of the Broad Street investigation and its maps. The great promise of urban density is that it thrusts so many diverse forms of intelligence, amateur and professional, into such a small space. The great challenge is figuring out a way to extract all that information and spread it throughout the community. The information that Snow and Whitehead sought revolved around the terror and senselessness of a deadly outbreak, but their approach has survived to tackle a vast array of problems, now augmented by modern information technology. Some of these problems are equally life-threatening ("When will my insulin go bad?"), but most of them involve the small concerns of everyday life. Add up enough of those small concerns, though, and you get a genuine transformation in your lived environment, not to mention a renewed sense of civic participation, a sense that your street-level understanding of your neighborhood can make a difference in the larger scheme of things. When Snow and Whitehead took their local knowledge of the Soho community and transformed it into a bird's-eye view of the outbreak, they were helping to invent a way of thinking about urban space whose possibilities we are still exploring today. It was an act with profound implications for the medical establishment, of course, but it was something else as well: a model for managing and sharing information that has implications that extend far beyond epidemiology.

The model involves two key principles, both of which are central to the way cities generate and transmit good ideas. First; the importance of amateurs and unofficial "local experts." Despite Snow's advanced medical training, the Broad Street case might well have been ultimately ruled in favor of miasma had it not been for the untrained local expertise of Henry Whitehead. Cities are invariably shaped by their master planners and their public officials; Chadwick and Farr had a tremendous impact on Victorian London—most of it positive, despite the miasma diversions. But in the last instance, the energy and vitality and innovation of cities comes from the Henry Whiteheads—the connectors and entrepreneurs and public characters who make the urban engine work at the street level. The beauty of technologies like 311 is that they amplify the voices of these local experts, and in doing so they make it easier for the authorities to learn from them.

The second principle is the lateral, cross-disciplinary flow of ideas. The public spaces and coffeehouses of classic urban centers are not organized into strict zones of expertise and interest, the way most universities or corporations are. They're places where various professions intermingle, where different people swap stories and ideas and skills along the way. Snow himself was a kind of one-man coffeehouse: one of the primary reasons he was able to cut through the fog of miasma was his multidisciplinary approach, as a practicing physician, mapmaker, inventor, chemist, demographer, and medical detective. But even with that polymath background, he still needed to draw upon an entirely different set of skills—more social than intellectual—in the form of Henry Whitehead's local knowledge.

WHEN SNOW PROPHESIED TO HIS FRIEND THAT THE TWO OF them might not live to see the waterborne theory vindicated, he was only half right. Snow died before his ideas could change the world, but Whitehead lived another four decades, long enough to see London fend off the Hamburg outbreak of 1892. Whitehead remained at St. Luke's until 1857, and then for the next seventeen years was a vicar at various parishes around the city, spending much of his time working on the problem of juvenile delinquency. In 1874, he left the city to serve on a series of ministries in northern England. Shortly before he left, his fellow investigator from the East End outbreak of 1866, John Netten Radcliffe, wrote of Whitehead's role in the Broad Street case:

> In the Broad Street outbreak of cholera not only did Mr. Whitehead faithfully discharge the duties of a parish priest, but by a subsequent inquiry, unique in character and extending over four months . . . he laid the first solid groundwork of the doctrine that cholera may be propagated through the medium of drinking water. . . . This doctrine, now fully accepted in medicine, was originally advanced by the late Dr. Snow; but to Mr. Whitehead unquestionably belongs the honour of having first shown with anything approaching to conclusiveness the high degree of probability attaching to it.

Henry Whitehead died in 1896, at the age of seventy. Until death, a portrait of his old friend John Snow hung in his study—to remind him, as Whitehead put it, "that in any profession the highest order of work is achieved, not by fussy empirical demands for 'something to be done,' but by patient study of the eternal laws."

How much would Henry Whitehead recognize were he to stroll down the streets of Soho today? The visible signs of the Broad Street outbreak would be long gone. Indeed, it is the peculiar nature of epidemic disease to create terrible urban carnage and leave almost no trace in the infrastructure of the city. The other great catastrophes that afflict cities—fires, earthquakes, hurricanes, bombs—almost invariably inflict vast architectural damage alongside the human body count. In fact, that's how they tend to do their killing: by destroying human shelter. Plagues are more insidious. The microbes don't care about buildings, because the buildings don't help them reproduce. So the buildings get to continue standing. It's the bodies that fall.

The buildings have changed nonetheless. Almost every structure that stood on Broad Street in the late summer of 1854 has been replaced by something new—thanks in part to the Luftwaffe, and in part to the creative destruction of booming urban real estate markets. (Even the street names have been altered. Broad Street was renamed Broadwick in 1936.) The pump, of course, is long gone, though a replica with a small plaque stands several blocks from the original site on Broad Street. A block east of where the pump once stood is a sleek glass office building designed by Richard Rogers with exposed piping painted a bold orange; its glassed-in lobby hosts a sleek, perennially crowded sushi restaurant. St. Luke's Church, demolished in 1936, has been replaced by the sixties development Kemp House, whose fourteen stories house a mixed-use blend of offices, flats, and shops. The entrance to the workhouse on Poland

Street is now a quotidian urban parking garage, though the workhouse structure is still intact, and visible from Dufours Place, lingering behind the postwar blandness of Broadwick Street like some grand Victorian fossil.

But there is much that Whitehead would recognize in the streets of Soho today, even though the buildings have been replaced and the rents have risen. The coffee shops are now mostly national chains, but the rest of the neighborhood is thick with the small-scale energy of local entrepreneurs. The mineral-teeth manufacturers have given way to video production facilities, hipster music stores with vinyl records in the window, Web design firms, boutique ad agencies, and "Cool Britannia" bistros—not to mention the occasional sex worker, a reminder of Soho's sordid days in the seventies. Everywhere the neighborhood is thriving with the passions and provocations of dense metropolitan living. The streets feel alive, precisely because they are animated by the intersecting paths of so many separate human lives. That there is safety and energy and possibility in those intersections—and not a looming fear of death—is part of the legacy of the battle fought on those streets 150 years ago. Perhaps it is even the most important part.

On Broad Street itself, only one business has remained constant over the century and a half that separates us from those terrible days in September 1854. You can still buy a pint of beer at the pub on the corner of Cambridge Street, not fifteen steps from the site of the pump that once nearly destroyed the neighborhood. Only the name of the pub has changed. It is now called The John Snow.

MARLBOROUGH STREET
STREET
NOEL
POLAND STREET
BERWICK
PORTLAND STREET
PORTLAND MEWS
WORK HOUSE
BENTINCK STREET
EDWARD ST
MARSHALL ST
WEST STREET
MARSHALL STREET
DUFOURS PLACE
BROAD STREET
PUMP
SOUTH ROW
CROSS ST
CAMBRIDGE ST
BREWERY
NEW STREET
HOPKINS STREET
COCK CT
HUBBARD ST
WILLIAM AND MARY YARD
LITTLE WINDMILL STREET
GREAT PULTENEY STREET
BRIDLE STREET
SILVER STREET
UP JAMES ST
UP JOHN ST
GOLDEN SQUARE
WARWICK STREET
PUMP
LOW JAMES ST
LOWER JOHN STR
LEICESTER ST
PUMP
STREET
SMITHS COURT
SMITHS YARD
WELLINGTON MEWS
LITTLE

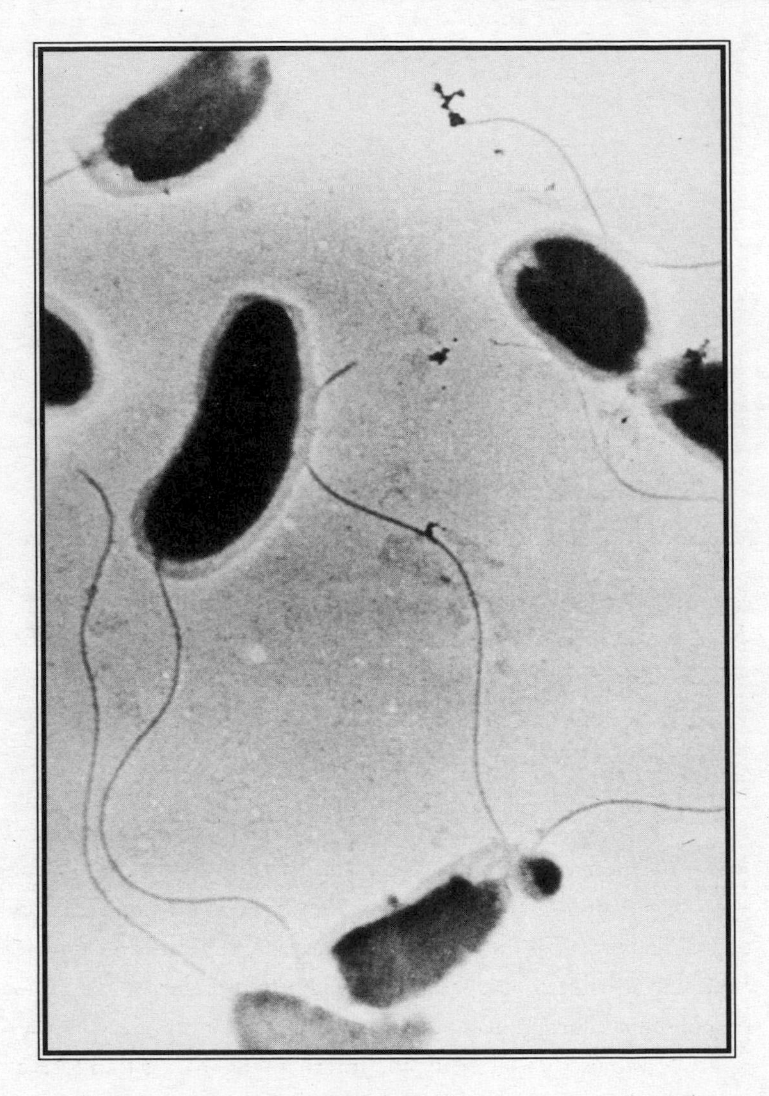

VIBRIO CHOLERAE

Epilogue

BROAD STREET REVISITED

SOMEWHERE IN THE WORLD, RIGHT ABOUT NOW, A VILLAGER is moving her family to a city somewhere, or an urban dweller is giving birth, or a farmer is dying—and with that local, isolated act, the global scales will tip decisively. We will enter a new era: a planet whose human population is more than 50 percent urban. Some experts believe we are on a path that will take us all the way to 80 percent, before we reach a planetary stabilization point. When John Snow and Henry Whitehead roamed the urban corridors of London 1854, less than 10 percent of the planet's population lived in cities, up from 3 percent at the start of the century. Less than two centuries later, the urbanites have become an absolute majority. No other development during that period—world wars, the spread of democracy, the use of electricity, the Internet—has had as transformative *and* widespread an impact on the lived experience of being human. The history books tend to orient themselves around nationalist story lines: overthrowing the king, electing the presidents, fighting the battles. But the history book of recent *Homo sapiens* as a species should begin and end with one narrative line: We became city dwellers.

If you time-traveled back to the London of September 1854 and described to some typical Londoners the demographic future that awaited their descendants, no doubt many would react with horror at the prospect of a "city planet," as Stewart Brand likes to call it. Nineteenth-century London was an overgrown, cancerous monster, doomed to implode sooner or later. Two million people crowded into a dense urban core was a kind of collective madness. Why would anyone want to do the same with *twenty* million?

To date, those fears have proved unfounded. Modern urbanization has thus far offered up more solutions than problems. Cities continue to be tremendous engines of wealth, innovation, and creativity, but in the 150 years that have passed since Snow and Whitehead watched the death carts make their rounds through Soho, they have become something else as well: engines of health. Two-thirds of women living in rural areas receive some kind of prenatal care, but in cities, the number is more than ninety percent. Nearly eighty percent of births in cities take place in hospitals or other medical institutions, as opposed to thirty-five percent in the countryside. For those reasons, as you move from rural areas to urban ones, infant mortality rates tend to drop. The vast majority of the world's most advanced hospitals reside in metropolitan centers. According to the coordinator of the United Nations Global Report on Human Settlements, "Urban areas offer a higher life expectancy and lower absolute poverty and can provide essential services more cheaply and on a larger scale than rural areas." For most of the world's nations, living in a city now extends your life expectancy instead of shortening it. Thanks to the government interventions of the seventies and eighties, air quality in many cities is as good as it has been since the dawn of industrialization.

Cities are a force for environmental health as well. This may be the most surprising new credo of green politics, which has in the past largely associated itself with a back-to-nature ethos that was explicitly antiurban in its values. Dense urban environments may do away with nature altogether—there are many vibrantly healthy neighborhoods in Paris or Manhattan that lack even a single tree—but they also perform the crucial service of reducing mankind's environmental footprint. Compare the sewage system of a midsized city like Portland, Oregon, with the kind of waste management resources that would be required to support the same population dispersed across the countryside. Portland's 500,000 inhabitants require two sewage treatment plants, connected by 2,000 miles of pipes. A rural population would require more than 100,000 septic tanks, and 7,000 miles of pipe. The rural waste system would be several times more expensive than the urban version. As the environmental scholar Toby Hemenway argues: "Virtually any service system—electricity, fuel, food—follows the same brutal mathematics of scale. A dispersed population requires more resources to serve it—and to connect it together—than a concentrated one." From an overall ecosystems perspective, if you're going to have 10 million human beings trying to share an environment with other life-forms, it's much better to crowd all 10 million of them into a hundred square miles than it is to spread them out, edge-city style, over a space ten or a hundred times that size. If we're going to survive as a planet with more than 6 billion people without destroying the complex balance of our natural ecosystems, the best way to do it is to crowd as many of those humans into metropolitan spaces and return the rest of the planet to Mother Nature.

By far, the most significant environmental cause that cities support is simple population control. People have more babies in the country, for a number of reasons. Economically, having more children makes sense in agrarian environments: more hands to help in the fields and around the house, without the space constraints of urban living. Rural life—particularly in the Third World—doesn't offer the same ready access to birth control and family-planning clinics. Cities, on the other hand, trend in the opposite direction, offering increased economic opportunity for women, expensive real estate, availability of birth control. Those incentives have turned out to be so powerful that they have reversed one of the dominant demographic trends of the last few centuries of life on earth: the population explosion that has been the subject of countless doomsday scenarios, from Malthus to Paul Ehrlich's influential early-1970s manifesto *The Population Bomb.* In countries that organized into modern metropolitan cities long ago, birthrates have dropped below the "replacement level" of 2.1 children per woman. Italy, Russia, Spain, Japan—all these countries are seeing birthrates around 1.5 per woman, which means that their populations will begin shrinking in the coming decades. The same trend is occurring in the Third World: birthrates were as high as 6 children per woman in the 1970s; now they are 2.9. As urbanization continues worldwide, current estimates project that the earth's human population will peak at around 8 billion in 2050. After that, it's a population *implosion* that we'll have to worry about.

THIS IS THE WORLD THAT SNOW AND WHITEHEAD HELPED make possible: a planet of cities. We no longer doubt that metropolitan centers with tens of millions of people can be a sustainable proposition, the way Victorian Londoners worried about the long-term viability of their sprawling, cancerous metropolis. In fact, the runaway growth of metropolitan centers may prove to be essential in establishing a sustainable future for humans on the planet. That reversal of fortune has much to do with the shifting relationship between microbe and metropolis that the Broad Street epidemic helped set in motion. "Cities were once the most helpless and devastated victims of disease, but they became great disease conquerors," Jane Jacobs wrote, in one of many classic passages from *Death and Life of the Great American City*.

> All the apparatus of surgery, hygiene, microbiology, chemistry, telecommunications, public health measures, teaching and research hospitals, ambulances and the like, which people not only in cities but also outside them depend upon for the unending war against premature mortality, are fundamentally products of big cities and would be inconceivable without big cities. The surplus wealth, the productivity, the close-grained juxtaposition of talents that permit society to support advances such as these are themselves products of our organization into cities, and especially into big and dense cities.

Perhaps the simplest way to explain why Broad Street was such a watershed event is to borrow Jacobs' phrase and say it this way: Broad Street marked the first time in history when a reasonable person might have surveyed the state of urban life and come to the conclusion that cities would someday become great conquerers of disease. Until then, it looked like a losing battle all the way.

Ultimately, the transformation that Broad Street helped usher in revolved around density, capitalizing on the advantages of dense urban living while minimizing the dangers. Crowding two hundred people per acre, building cities with populations in the millions sharing the same water supply, struggling to find a way to get rid of all that human and animal waste—this was a lifestyle decision that seemed fundamentally at odds with both personal and environmental health. But the nations that first organized themselves around metropolitan settlements—as turbulent as those transformations were—are now the most affluent places on the planet, with life expectancies that are nearly double that of predominantly rural nations. A hundred and fifty years after Broad Street, we see density as a positive force: an engine of wealth creation, population reduction, environmental sustainability. We are now, as a species, dependent on dense urban living as a survival strategy.

But the forecasts that predict a city-planet where eighty percent of us live in metropolitan areas are just that: forecasts. It is possible that this epic transformation could be undone in the coming decades or centuries. The rise of sustainable metropolitan environments was not a historical inevitability: it was the result of specific technological, institutional, economic, and scientific developments, many of which played a role in the extended story of Broad Street. It's entirely possible that new forces could emerge—or old foes return—that would imperil this city-planet of ours. But what might they be?

It is unlikely that these antiurban forces will come in the form of some new incentive that lures people back to the countryside, like the fanciful dream of telecommuting prophesied by the futurists a decade ago, when the Internet was first entering mainstream culture.

There's a reason why the world's wealthiest people—people with near-infinite options vis-à-vis the choice of where to make their home—consistently choose to live in the densest areas on the planet. Ultimately, they live in these spaces for the same reason that the squatter classes of São Paulo do: because cities are where the action is. Cities are centers of opportunity, tolerance, wealth creation, social networking, health, population control, and creativity. No doubt, the Internet and its descendants will continue exporting some of these values to rural communities in the decades to come. But of course, the Internet will continue enhancing the experience of urban life as well. The sidewalk flaneurs get as much out of the Web as the ranchers do, if not more.

The two great looming threats of our new century—global warming and our finite supply of fossil fuels—may well have massively disruptive effects on existing cities in the coming decades. But they are not likely to disrupt the macro pattern of urbanization in the long run, unless you believe the environmental crisis is likely to end in some global cataclysm that sends us back to agrarian or hunter-gatherer living. Most of the world's urban centers lie within a few dozen meters of sea level, and if the ice caps do indeed melt at the rate they are currently forecast to, many of our metropolitan descendants will be relocating by the midsection of the twenty-first century. But there's no reason to think they'll be relocating to rural or suburban areas. Most likely, they'll simply retreat to higher ground, and new dense metropolitan areas will form around them. The wealthiest cities of the world will follow Venice's lead and simply try to engineer their way around the problem. The poorest cities will follow New Orleans' lead—at least so far—and just move to other nearby cities. Either way the poplation stays urban.

Neither does the end of oil foretell the end of cities. The reason why cities have taken on the "green" stamp of approval in recent years is not that they are literally green with foliage. (Air quality has improved markedly, and parks are better funded than ever, but they remain concrete jungles for the most part.) We now see cities as environmentally responsible communities because their energy footprints are so much smaller than other forms of human settlement. In a sense, the environmentalists are learning something that the capitalists learned a few centuries ago: There are efficiencies to urban living that outweigh all the annoyances. City dwellers spend less money heating and cooling their homes; they have fewer children; they recycle their waste more economically; and most important, they consume less energy moving around day to day, thanks to the shorter commutes and mass transit that density enables. "By the most significant measures, New York is the greenest community in the United States, and one of the greenest cities in the world," *The New Yorker*'s David Owen writes. "The most devastating damage humans have done to the environment has arisen from the heedless burning of fossil fuels, a category in which New Yorkers are practically prehistoric. The average Manhattanite consumes gasoline at a rate that the country as a whole hasn't matched since the mid–nineteen-twenties, when the most widely owned car in the United States was the Ford Model T. Eighty-two per cent of Manhattan residents travel to work by public transit, by bicycle, or on foot. That's ten times the rate for Americans in general, and eight times the rate for residents of Los Angeles County. New York City is more populous than all but eleven states; if it were granted statehood, it would rank fifty-first in per-capita energy use." In other words, a serious crisis of nonrenewable energy resources is likely to *accelerate* the urbanization trend, not derail it.

None of this is intended to belittle the long-term problems

caused by global warming and our dependence on fossil fuels. Both trends are likely to trigger disastrous consequences if left unchecked, and the sooner we get serious about solutions to both problems, the better. But in both cases, one of the primary solutions may well prove to be to encourage people to move to metropolitan areas. A warmer planet is still a city-planet, for better or worse.

Yet that doesn't mean continued urbanization is inevitable. It just means that the potential threats will come from somewhere else. Most likely, if some new force derails our mass migration to the cities, it will take the form of a threat that specifically exploits density to harm us, just as *Vibrio cholerae* did two hundred years ago.

In the immediate aftermath of the 9/11 attacks, many commentators observed that there was a certain dark irony in the technological method of the terrorists: they had used what were effectively Stone Age tools—knives—to gain control of advanced American machines—four Boeing 7-series planes—and then employed that technology as a weapon against its creators. But while the planes were clearly instrumental to the attack, the advanced technology that caused the greatest loss of life lay elsewhere: the terrorists also exploited the technical knowledge that enabled 25,000 people to occupy a building 110 stories high. (Consider that a dead-on collision with the five-story Pentagon produced only seventy-nine casualties on the ground.) The heat of jet fuel and the impact of a 400-mph collision were lethal weapons that morning, but without the terrifying potential energy released by those collapsing floors, the body count would have been lower by an order of magnitude.

The 9/11 attackers were, ultimately, exploiting the tremendous advance in the technologies of density that we have enjoyed since the birth of skyscrapers in the late nineteenth century. There were four hundred people per acre in Soho in 1854, in London's most densely populated neighborhood. The Twin Towers sat on approximately one acre of real estate, and yet they harbored a population of 50,000 on a workday. That level of density offers a long list of potential benefits, but it is also an open invitation for mass killing—and, what's worse, mass killing that doesn't require an army to carry it out. You just need enough ammunition to destroy two buildings, and right there you've got a body count that rivals the ten years of American losses in the Vietnam War.

Density is the crucial ingredient often left out in discussions of asymmetric warfare. It is not merely that technology has given smaller and smaller organizations access to increasingly deadly weapons—though that is surely half the story—but that the patterns of human settlement over the past two hundred years have made those weapons far more deadly than they would be if one could somehow time-travel back to 1800 and set them off. Even if you could have hijacked an airplane back in John Snow's day, you'd have been hard pressed to find an urban area crowded enough to kill a hundred civilians on the ground. Today, the planet is covered with thousands of cities that offer far more enticing targets. If terrorist-sponsored asymmetric warfare were the only threat facing human beings, we would be far better off as a species covering the planet with suburban sprawl and emptying the cities altogether. But we don't have that option. So we're either going to have to acclimate to a certain predictable presence of terrorist threats—the way the Victorian Londoners acclimated to the terrible plagues that would sweep through their city every few years—or we're going to have to follow John Snow's lead and figure out a reliable way to eliminate the threat.

Certain threats, however, may not be tolerable. One of the most menacing that the twenty-first-century city faces is a holdover from the Cold War: nuclear weapons. The doomsday scenarios are familiar enough: A megaton hydrogen bomb—too big for "suitcase bombs" but much smaller than today's twenty-five-megaton state-of-the-art weapons—detonated at the site of the Broad Street pump vaporizes the entire area from the western edge of Hyde Park to Waterloo Bridge. A weekday attack would effectively wipe out the entire British government, reducing the Houses of Parliament and 10 Downing Street to radioactive ash. Most of London's landmarks—Buckingham Palace, Big Ben, Westminster Abbey—would simply cease to exist. A wider zone extending out to Chelsea and Kensington and to the eastern edge of the old City would suffer 98 percent loss of life. Move a few miles farther out—up to Camden Town, out to Notting Hill or the East End—and half the population dies, with most buildings damaged beyond recognition. Anyone who happens to see the blast directly is blinded for life; most survivors suffer hideous radiation sickness that makes them envy the dead. As you move out from Ground Zero, the fallout leaves a vast wake of elevated cancer occurrences and genetic defects.

Then there are the secondary effects, the collateral damage. The entire government would have to be replaced overnight; the damage to the financial centers in the city would be catastrophic for the world economy. The detonation site itself would be uninhabitable for decades. Every resident of a major world city—every New Yorker and Parisian, every person in every street in Tokyo and Hong Kong—would find his habitat transformed: from safety in numbers to mass terror. The great cities of the world would start to look like giant bull's-eyes: millions of potential casualties conveniently stacked up in easily demolished high-rises. One such attack would probably not impede the metropolitan migration—after all, Hiroshima and

Nagasaki didn't stop Tokyo from becoming the world's largest city. But several detonations might well tip the balance. Turn our metropolitan centers into genuine nuclear targets and you risk a whole other kind of "nuclear winter": a season of mass exodus unrivaled in human history.

It would be bad news, in other words. And this bad news is likely to arrive courtesy of a walk-on part on the world-historical stage, somebody driving a rigged SUV into Soho and pulling the trigger. There are 20,000 nuclear weapons in the world capable of inflicting this level of damage. That we know about. On a planet of more than 6 billion people, there have to be thousands and thousands of lost souls ready and willing to detonate one of those weapons in a crowded urban center. How long before those two sets intersect?

That driver with the rigged SUV isn't going to be deterred by the conventional logic of détente-era nuclear politics. Mutually assured destruction isn't much of a deterrent to him. Mutually assured destruction, in fact, sounds like a pretty good outcome. Game theory has always had trouble accounting for players with no rational self-interest, and the theories of nuclear deterrence are no exception. And once the bomb goes off, there's no second line of defense—no vaccines or quarantines to block off the worst-case scenario. There will be maps, but they'll be maps of incineration and fallout and mass graves. They won't help us understand the threat the way Snow's map helped us understand cholera. They will merely document the extent of the tragedy.

THE PERILS OF DENSITY GROW MORE EXPLOSIVE—OR MORE infectious, as the case may be—as the wages of fear are increasingly doled out in twenty-first-century currency: chemical or biological weapons, a freelancer virus or bacterium terrorizing the planet for no particular cause other than its fundamental drive to reproduce. When people still worry about the long-term sustainability of dense human settlement, it is more often than not these self-replicating weapons that conjure up the doomsday scenarios. Tightly bound networks of humans and microbes make a great case study in the power of exponential growth. Infect ten people with the *Ebola* virus in Montana and you might end up killing a hundred others, depending on when the initial victims were taken to the high-density environment of a hospital. But infect ten people with *Ebola* in downtown Manhattan and you could kill a million, or more. Traditional bombs obviously grow more deadly as the populations they target increase in size, but the upward slope in that case is linear. With epidemics, the deadliness grows exponentially.

In September 2004, health officials in Thailand began a program of vaccinating poultry workers with the conventional flu shots that are routinely doled out in Western countries at the start of flu season every year. For months, health experts around the world had been calling for precisely this intervention. This, in itself, was a telling phenomenon. Conventional flu vaccines are effective against only the type A and type B strains of influenza—the kind that sidelines you for a week with a fever and a stuffy head, but that is rarely fatal in anyone except the very young or the very old. The risk of a global pandemic emerging from these viruses is slim at best, which is why, historically, public-health officials in the West have not concerned themselves with the question of whether poultry workers on the other side of the world have received their flu shots. The virus that the public-health officials *were* worried about—H5N1, also known as the avian flu—is entirely unfazed by conventional flu shots. So why were so many global health organizations calling for vaccines in

Asia? If they were worried about avian flu, why prescribe a vaccine that was known to be ineffective against it?

The answer to that question is a measure of how far we have come since the Broad Street epidemic in our understanding both of the pathways that disease takes and the underlying genetic code that instructs bacteria and viruses. But it is also a measure of continuity: how the very same issues that Snow and Whitehead confronted on the streets of London have returned to haunt us, this time on the scale of the globe and not the city. The specific threats are different now, and in some ways they are more perilous, and the tools at our disposal are far more advanced than Snow's statistical acumen and shoe-leather detective work. But confronting these threats requires the same kind of thinking and engagement that Snow and Whitehead so brilliantly applied to the Broad Street outbreak.

In all the speech-making, posturing, and sober analysis about avian flu that has swept the globe in the past decade, one utterly amazing fact stands out: as far as we know, the virus that has caused such international panic *does not exist yet.* To be sure, H5N1 is a viciously lethal virus, with fatality rates in humans approaching 75 percent. But in its current incarnation, it is incapable of starting a pandemic, because it lacks the ability to pass directly from human to human. It can spread like wildfire through a population of chickens or ducks, and the birds can in turn infect humans. But there the chain of infection ends: so long as the overwhelming majority of humans on the planet are not in direct contact with live poultry, H5N1 is incapable of causing a global outbreak.

So why are health officials in London and Washington and Rome worried about poultry workers in Thailand? Why, indeed, are these officials worried about avian flu in the first place? Because microbial life has an uncanny knack for mutation and innovation. All the world needs is for a single strain of H5N1 to somehow mutate into a form that *is* transmissible between humans, and that virus could unleash a pandemic that could easily rival the 1918 influenza pandemic, which killed as many as 100 million people worldwide.

That new capability might come from some random mutation in the H5N1 DNA. For the H5N1, it would be like winning a genetic lottery where the odds were a trillion-to-one against you, but in a world with untold trillions of H5N1 viruses floating around, it's not impossible to imagine. But the more likely scenario is that H5N1 will borrow the relevant genetic code directly from another organism, in a process known as transgenic shift. Recall that DNA transmission among single-celled bacteria and viruses is far more promiscuous than the controlled, vertical descent of all multicellular life. A virus can swap genes with other viruses willingly. Imagine a brunette waking up one morning with a shock of red hair, after working side by side with a redheaded colleague for a year. One day the genes for red hair just happened to jump across the cubicle and express themselves in a new body. It sounds preposterous because we're so used to the way DNA works among the eukaryotes, but it would be an ordinary event in the microcosmos of bacterial and viral life.

Most conventional flu viruses already possess the genetic information that allows them to pass directly from human to human. Because H5N1 is so closely related to the conventional flu virus, it would be a relatively simple matter for it to swipe a few lines of pertinent code and immediately enjoy its new capacity for human-to-human transmission. Certainly it would be easier than randomly stumbling on the correct sequence via mutation.

And so this is why the whole world has suddenly taken an interest in whether Thai poultry workers get their flu shots: because the world wants to ensure that H5N1 stays as far away as possible from ordinary flu viruses. If the two viruses did encounter each other inside a human host, a far more ominous strain of H5N1 might emerge. It could be as infectious as the influenza bug that swept the globe in 1918, but several times more lethal. And it would find itself inhabiting a planet that was massively more interconnected and densely settled than it was in 1918.

To appreciate how deadly transgenic shift can be, you need only look at the Broad Street epidemic. In 1996, two scientists at Harvard, John Mekalanos and Matthew K. Waldor, made an astonishing discovery about the roots of *Vibrio cholerae*'s killer instinct. There are two key components to the bacteria's assault on a human body: the TCP pilus that allows it to replicate with such exponential fury in the small intestine, and the cholera toxin that actually triggers the rapid dehydration of the host. Mekalanos and Waldor discovered that the gene for cholera toxin is actually supplied by an outside source: a virus called CTX phage. Without the genes contributed by that virus, *V. cholerae* literally doesn't know how to be a pathogen. It learns to be a killer by borrowing genetic information from an entirely different species. The trade between the phage and the bacterium is a classic example of coevolutionary development, two organisms cooperating at the genetic level in order to further both of their reproductive interests: the CTX phage multiplies inside the *V. cholerae,* and in return the virus offers up a gift that allows the bacteria to greatly increase the odds of finding another host to infect. As unlikely as it sounds, *V. cholerae* is not a born killer. It needs the CTX phage to switch over to the dark side.

So we have good reason to fear genetic commingling between H5NI and the ordinary human flu virus. But we should also be comforted by how far we have advanced in our ability to anticipate these cross-species transmissions. When John Snow identified the waterborne nature of cholera in the middle of the nineteenth century, he was using the tools of science and statistics to find a way around the fundamental perceptual limits of space: the creature he was seeking was literally too small to see. So he had to detect it indirectly: in patterns of lives and deaths that played out in the streets and houses of a bustling metropolitan center. Today we have conquered that spatial dimension: we can visually inspect the kingdom of bacteria at will; we can even zoom all the way down to the molecular strands of DNA, even glimpse the atomic connections that bind them together. So now we confront another fundamental perceptual limit—not of space, but of time. We use the same methodological tools that Snow used, only now we're using them to track a virus we can't see because it doesn't exist yet. Those flu vaccinations in Thailand are a preemptive strike against a possible future. No one knows when H5N1 will learn to pass directly from human to human, and it remains at least a theoretical possibility that it will never develop that trait. But planning for its emergence makes sense, because if such a strain does appear and starts spreading around the globe, there won't be the equivalent of a pump handle to remove.

This is why we're vaccinating poultry workers in Thailand, why the news of some errant bird migration in Turkey can cause shudders in Los Angeles. This is why the pattern recognition and local knowledge and disease mapping that helped make Broad Street understandable have never been more essential. This is why a continued commitment to public-health institutions remains one of the most vital roles of states and international bodies. If H5N1 does manage to swap just the right piece of DNA from a type A flu virus, we could well see a runaway epidemic that would burn through some of the world's largest cities at a staggering rate, thanks both to the extreme densities of our cities and the global connectivity of jet travel.

Millions could die in a matter of months. Some experts think a pandemic on the order of 1918 is a near inevitability. Would a hundred million dead—the great majority of them big-city dwellers—be enough to derail the urbanization of the planet? It's unlikely, as long as new pandemics didn't start rolling in every flu season, like hurricanes. But think of the lingering trauma that 9/11 inflicted on every New Yorker—wondering if it was still safe to stay in the city. Almost everyone opted to stay, of course, and the city's population has—wonderfully—continued to swell, thanks largely to immigration from the developing world.

But imagine if 500,000 New Yorkers had died of the flu in September 2001, instead of 2,500 in a collapsing skyscraper. Just the deaths alone would give the year the ignominious status of the single most dramatic drop in population in the city's history, and no doubt the deaths would be exceeded by all the migrations to the relative safety of the countryside. My wife and I are passionately committed to the idea of raising our kids in an urban environment, but if 500,000 New Yorkers were killed in the space of a few months, I know we'd find another home. We'd do it with great regret, and with the hope that, when things settled down a few years later we'd move back. But we would move, all the same.

It is conceivable, then, that a living organism—whether the product of evolution or genetic engineering—could threaten our great transformation into a city-planet. But there is reason for hope. We have a window of a few decades where DNA-based microbes will retain the capability of unleashing a cascading epidemic that kills a significant portion of humanity. But at a certain point—perhaps ten years from now, perhaps fifty—the window may well close, and the threat may subside, just as other, more specific, biological threats have subsided in the past: polio, smallpox, chicken pox.

If this scenario comes to pass, the pandemic threat will ultimately be defeated by a different kind of map—not maps of lives and deaths on a city street, or bird flu outbreaks, but maps of nucleotides wrapped in a double helix. Our ability to analyze the genetic composition of any life-form has made astonishing progress over the past ten years, but in many ways we are at the very beginning of the genomic revolution. We have already seen amazing advances in our *understanding* of the way genes build organisms, but the *application* of that understanding—particularly in the realm of medicine—is only starting to bear fruit. A decade or two from now, we may possess tools that will allow us to both analyze the genetic composition of a newly discovered bacterium and, using computer modeling, build an effective vaccine or antiviral drug in a matter of days. At that point, the primary issue will be production and delivery of the drugs. We'll know how to make a cure for any rogue virus that shows up; the question will be whether we can produce enough supplies of the cure to stop the path of the disease. That might well require a new kind of urban infrastructure, a twenty-first-century equivalent of Bazalgette's sewers: production plants located in every metropolitan center, ready to churn out millions of vaccines if an epidemic appears. It will take the creation of public-health institutions in the developing world—institutions that simply do not exist yet—along with a renewed commitment to public health in the industrialized world, particularly the United States. But we'll have the tools at our disposal to deal with the emerging threats, if we're smart enough to deploy them.

The twentieth-century approach to battling viruses has largely operated at the same temporal scale as microbial evolution itself. It has been a classic Darwinian arms race. We take a sample of last year's most prolific flu virus and use it as the basis for a vaccine that we then spread through the immune system of the general public; and the viruses evolve new ways around those vaccines, and so we come up with new vaccines that we hope will deal with the new bugs. But the genomic revolution means that our defense mechanisms are now starting to operate at a much faster clip than evolution. We're no longer limited to jury-rigging vaccines out of last year's model. We're able to project forward, anticipate future variations, and, increasingly, address the specific threat posed by the most active virus on the ground. Our understanding of the building blocks of life is advancing at nearly exponential rates—thanks in part to the exponential advance in computation power we call Moore's Law. But the building blocks themselves are not getting more complex. Type A influenza possesses only eight genes. Thanks to the transgenic shift of microbial life, those eight genes are capable of an astonishing amount of variation; but those possibilities are ultimately finite, and they will be no match for the modeling prowess of circa-2025 technology. Right now we're in an arms race with the microbes, because, effectively, we're operating on the same scale that they are. The viruses are both our enemy and our arms manufacturer. But as we enter an age of rapid molecular analysis and prototyping, the whole approach changes. The complexity of our understanding of microbial diseases is already advancing much faster than the complexity of the microbes themselves. Sooner or later, the microbes won't be able to compete.

But perhaps the arms race will not purely be a figure of speech. The flu virus on its own might not be able to grow complex enough to challenge the technology of genomic science, but what if the technology of genomic science were used to "weaponize" a virus? Genetic engineering may ultimately win out over evolution, but isn't it a different matter if the viruses are themselves the product of genetic engineering? Wouldn't the ominous trends of asymmetric warfare—increasingly advanced technology in the hands of smaller and smaller groups—be even more ominous where biological weapons are concerned? If suicide bombers with homemade explosives can effectively hold the American military hostage, imagine what they could do with a weaponized virus.

The crucial difference, though, is that there are vaccines for biological weapons, while there are no vaccines for explosives. Any DNA-based agent can effectively be neutralized after its release, by any number of different mechanisms: early detection and mapping, quarantine, rapid vaccination, antiviral drugs. But you can't neutralize an explosive once it has been detonated. So suicide bombers are probably destined to be a part of human civilization for as long as there are political or religious ideologies that encourage people to blow themselves up in crowded places. DNA-based weapons do not have the same future, however, because for every terrorist trying to engineer a biological weapon there are a thousand researchers working on a cure. It's entirely likely, of course, that we will see the release of an infectious agent engineered in a rogue lab somewhere, and it's at least conceivable that the attack could unleash a pandemic that could kill thousands or millions—particularly if such an attack took place in the next decade or so, before our defensive tools have matured. But there's good reason to believe that defensive tools will ultimately win out in this domain as well, because they will be built on a meta-understanding of genetics itself, and because the resources put into their development will dramatically outnumber the resources devoted to developing weapons—assuming, that is, that the world's nation-states continue the ban on the creation of biological weapons. Biological terrorism may well be in our future, and it could turn out to be one of the most hideous chapters in the history of human warfare. But in the long run, it shouldn't threaten our transformation into a city-planet, if we continue to encourage scientific research into defensive vaccines and other treatment, and remain vigilant in our opposition to state-sponsored biological weapons research.

Here, too, the legacy of Snow's map is essential to the battle. The peculiar menace of a biological attack is that we may not know it is under way until weeks after the infectious agent is first released. The greatest risk of a deliberately planned urban epidemic is not that we won't have a vaccine, it's that we won't recognize the outbreak until it's too late for the vaccine to stop the spread of disease. Combating this new reality will take a twenty-first-century version of John Snow's map: making visible patterns in the daily flow of lives and deaths that constitute the metabolism of a city, the rising and falling fortunes of the sick and the healthy. We'll have exceptional tools at our disposal to defend ourselves against a biological attack, but we'll have to be able to *see* the attack first, before we can apply those defensive measures. Before we can mobilize all the technology that would have bewildered Snow—the genomic sequencers and antiviral mass-production facilities—we'll use a technology that Snow would have recognized instantly. We'll use a map. Only, this map won't be hand-illustrated from data collected via door-to-door surveys. It will draw on the elaborate network of sensors sniffing the air for potential threats in urban centers, or hospital first-responders reporting unusual symptoms in their patients, or public water facilities scanning for signs of contamination. Almost two centuries after William Farr first hit upon the idea of amassing weekly statistics on the mortality of the British population, the technique he pioneered has advanced to a level of precision and scope that would have astonished him. The Victorians could barely see microbial life-forms swimming in a petri dish in front of them. Today, a suspicious molecule floats by a sensor in Las Vegas, and within hours the authorities at the CDC in Atlanta are on the case.

There is less reason for optimism where nuclear weapons are concerned. A technique that effectively neutralizes the threat posed by influenza viruses could come from any number of active lines of research: from our understanding of the virus itself, from our understanding of the human immune system, even our understanding of how the respiratory system works. There are thousands of scientists and billions of dollars spent every year exploring new ways to fight lethal epidemic diseases. But no one is working on a way to neutralize a nuclear explosion, presumably for the entirely rational reason that it is impossible to neutralize a nuclear explosion. We have made some advances in detection—all nuclear devices give off a radioactive signal that can be tracked by sensors—but detection is hardly a fail-safe option. (If we were relying purely on our ability to detect emerging viruses, the long-term future for epidemic disease would look equally grim.) There is some promising research into medicines that would block the effects of radiation poisoning, which could well save millions of lives in the event of a metropolitan detonation, but millions more would still perish from the initial explosion itself.

If you look solely at the danger side of the equation, both epidemic disease and nuclear explosions seem to present a mounting threat in the coming decades: thanks to urban density and global jet travel, it's probably easier now for a rogue virus to spread around the globe, while the breakup of the Soviet Union and the increase in technological expertise has made it easier to both acquire radioactive materials and build the bomb itself. (As I write, the world is wrestling with the implications of Iran's renewed commitment to a nuclear program.) But if you look at the opposing side of the equation—our ability to neutralize the threat—the story is very different. Our ability to render a virus harmless is growing at exponential rates, while our ability to undo the damage caused by the detonation of a nuclear device is, literally, nonexistent, with no sign that it will *ever* be technically possible.

On some level, the nuclear problem may turn out to be one that we never solve, and the ultimate question will turn out to be how often a rogue nation or terrorist cell manages to get its hands on one of these devices. Perhaps urban nuclear explosions will turn out to be like hundred-year storms: a bomb goes off once a century, millions die, the planet shudders in horror, and slowly goes about its business. If that's the pace, then as horrible as such a catastrophe would be, the long-term sense of urban sustainability would likely remain intact. But if the trends of asymmetric warfare continue, and the suicide bombers start detonating suitcase nukes every ten years—at that point, all bets are off.

AND SO OUR CONVERSION TO A CITY-PLANET IS BY NO means irreversible. The very forces that propelled the urban revolution in the first place—the scale and connectedness of dense urban living—could be turned against us. Rogue viruses or weapons could once again turn urban areas into sites of mass death and terror. But if we are to keep alive the model of sustainable metropolitan life that Snow and Whitehead helped make possible 150 years ago, it is incumbent on us to do, at the very least, two things. The first is to embrace—as a matter of philosophy and public policy—the insights of science, in particular the fields that descend from the great Darwinian revolution that began only a matter of years after Snow's death: genetics, evolutionary theory, environmental science. Our safety depends on being able to predict the evolutionary path that viruses and bacteria will take in the coming decades, just as safety in Snow's day depended on the rational application of the scientific method to public-health matters. Superstition, then and now, is not just a threat to the truth. It's also a threat to national security.

The second is to commit ourselves anew to the kinds of public-health systems that developed in the wake of the Broad Street outbreak, both in the developed world and the developing: clean water supplies, sanitary waste-removal and recycling systems, early vaccination programs, disease detection and mapping programs. Cholera demonstrated that the nineteenth-century world was more connected than ever before; that local public-health problems could quickly reverberate around the globe. In an age of megacities and jet travel, that connectedness is even more pronounced, for better and for worse.

In many ways the story of the past few years is not an uplifting one, where these two objectives are concerned. Intelligent design "theory" continues to challenge the Darwinian model, in the courts and in public opinion; the United States appears to be spending more time and money proposing new nuclear weapons than eliminating the ones we have; public-health spending is down per capita; as I write, Angola is suffering through the worst outbreak of cholera in a decade.

But if our current prospects seem bleak, we need only think of Snow and Whitehead on the streets of London so many years ago. The scourge of cholera then seemed intractable, too, and superstition seemed destined to rule the day. But in the end, or at least as close to the end as we've gotten so far, the forces of reason won out. The pump handle was removed; the map was drawn; the miasma theory was put to rest; the sewers were built; the water ran clean. This is the ultimate solace that the Broad Street outbreak offers us in our current predicament, with all its unique challenges. However profound the threats are that confront us today, they are solvable, if we acknowledge the underlying problem, if we listen to science and not superstition, if we keep a channel open for dissenting voices that might actually have real answers. The global challenges that we face are not necessarily an apocalyptic crisis of capitalism or mankind's hubris finally clashing with the balanced spirit of Gaia. We have confronted equally appalling crises before. The only question is whether we can steer around these crises without killing ten million people, or more. So let's get on with it.

AUTHOR'S NOTE

THIS BOOK IS A HISTORICAL NARRATIVE OF THE EVENTS OF September 1854 in London based on the many surviving eyewitness accounts and the exhaustive investigations by the authorities in the months after the outbreak subsided. Any direct dialogue quoted in the text comes from those firsthand accounts, and where ambiguities exist about names or the timing of events, I have made a note of it in the text or in the endnotes. The one literary convention that I have adopted is to attribute thoughts to some of the individuals at specific points in the narrative. In each case, the historical record is clear that the thought did occur to them at some point during the outbreak and its aftermath; I have simply made an educated guess as to when exactly the thoughts first came to mind.

ACKNOWLEDGMENTS

It occurred to me somewhere in the middle of writing *The Ghost Map* that this was a book I'd been preparing for almost twenty years, ever since I decided to do my undergraduate thesis on the way cultures respond to epidemics. In grad school a few years later, my primary focus was the metropolitan novel in Victorian society, specifically the imaginative challenge that confronted anyone who tried to represent the overwhelming experience that was London in that period. To the professors and friends who guided me then—Robert Scholes, Neil Lazarus, Franco Moretti, Steven Marcus, and the late Edward Said—thank you for steering me toward Broad Street with such intelligence and patience.

I'm indebted to a number of people who read the manuscript and improved the book immensely with their thoughts and corrections: Carl Zimmer, Paul Miller, Howard Brody, Nigel Paneth, Peter Vinten-Johansen, and Tom Koch. A number of scholars were kind enough to comment on specific sections of the manuscript, or to answer my questions about the material: Sherwin Nuland, Steven Pinker, Ralph Frerichs, John Mekalanos,

Sallie Patel, and Stewart Brand. My research assistant, Ivan Askwith, was once again an invaluable collaborator, as was Russell Davies, who came through with some last-minute additions from the streets (and libraries) of London. Whatever errors remain are mine alone.

I'm grateful to the many libraries whose resources I drew on in my research: those of Harvard, MIT, and NYU, and the New York Public Library. I am particularly indebted to two London institutions: the Wellcome Library for the History and Understanding of Medicine and, of course, the peerless British Library—even the remote newspaper reading rooms in Colindale. My editors at *Wired* and *Discover*—Steve Petranek, Dave Grogan, Chris Anderson, Ted Greenwald, Chris Baker, Mark Robinson, and Rob Levine—helped me explore, over the past few years, a number of the themes addressed in the closing chapters here. I'm also grateful for the friends who have made London such a wonderful place to visit, and who inspired me to write a book about the city in the first place: Hugh Warrender, Richard Rogers, Ruthie Rogers, Roo Rogers, Brian Eno, Helen Conford, and Stefan McGrath.

At Riverhead, I'm grateful for the support from the publicity team—Kim Marsar, Matthew Venzon, and Julia Fleischaker—who helped me survive the madness of *Everything Bad*'s media frenzy while I was writing this book. Thanks to Larissa Dooley for being on top of a million different threads at the same time. And thanks to my fearless editor, Sean McDonald, who sets some kind of record by being the first editor to make it through two books of mine. As for my agent, Lydia Wills—I got all gushy in the last acknowledgments and it's really gone to her head since then, so I'm not mentioning her at all this time around.

But as always, the acknowledgments begin and end with my wife, Alexa—the closest of readers—and our three boys: Clay, Rowan, and the latest addition, born not five days ago as I write, Dean.

Brooklyn
July 2006

APPENDIX:
NOTES ON FURTHER READING

THERE ARE TWO INDISPENSABLE RESOURCES FOR UNDERstanding the life and work of John Snow. The first is the exhaustive Web archive devoted to all things Snow, maintained by the UCLA epidemiology professor Ralph Frerichs. The site, accessible at www.ph.ucla.edu/epi/snow.html, has everything from annotated reproductions of various maps of the period to a multimedia tour of the Broad Street outbreak to a complete digital collection of Snow's writing. The second is *Cholera, Chloroform, and the Science of Medicine,* written by a multidisciplinary team of scholars (Peter Vinten-Johansen and others) from Michigan State University. The book is both a biography of Snow himself and a clear and insightful survey of the intellectual landscape he traveled during the course of his life. Both resources were essential to the writing of this book, and I highly recommend them for anyone interested in exploring John Snow's work in more detail.

For readers interested in the map itself, and in Snow's legacy as an information designer, Edward Tufte's account is by now the canonical one, though his initial telling of the story—in his 1983 book *The Visual Display of Quantitative Information*—was factually wrong on several fronts, as he acknowledged in his subsequent work, *Visual Explanations,* which offered a more nuanced account of the Broad Street outbreak (and which managed to reproduce Snow's map itself, instead of the secondhand copy that ran in the first book). Tom Koch's brilliant *Cartographies of Disease* offers a comprehensive look at Snow's place in the specific tradition of disease mapping.

There are innumerable portraits of Victorian London, but Henry Mayhew's *London Labour and the London Poor* is still the most riveting and thorough account of the city's vast underclass, rivaled only by Engels' London chapters from *The Condition of the English Working Class.* Among the contemporary accounts, Liza Picard's *Victorian London,* Roy Porter's *London: A Social History,* and Peter Ackroyd's *London: A Biography* are all worth reading. On the future of cities, I recommend Stewart Brand's essay "City Planet" and Richard Rogers' *Cities for a Small Planet.* The best account of the psychological and cultural impact of urbanization remains Raymond Williams' masterly *The Country and the City.* Stephen Halliday's *The Great Stink* tells the amazing story of Joseph Bazalgette's battle to build London's sewer system. For a modern look at waste management, I recommend William Rathje and Cullen Murphy's *Rubbish: The Archaeology of Garbage.* Readers interested in the social history of beverages—including tea, coffee, and spirits—will want to read Tom Standage's *A History of the World in Six Glasses.*

On the scale of bacteria, the seminal work in the field remains Lynn Margulis and Dorion Sagan's mind-opening *Microcosmos.* Though it doesn't deal directly with cholera, Carl Zimmer's *Parasite*

Rex is also a fascinating exploration of our microscopic fellow-travelers. For a unnerving look at the failure of modern public-health infrastructure, see Laurie Garrett's *Betrayal of Trust*.

The story of the Broad Street outbreak itself has been sketched in numerous books, usually with significant distortions. Many accounts assume that Snow created the map during the outbreak, or that he developed the waterborne theory from his investigations at Broad Street. Henry Whitehead is often ignored altogether. And so the best sources for understanding the outbreak are still John Snow and Henry Whitehead themselves. Their various published accounts of the events are available online at the UCLA site, and at a special John Snow archive hosted by Michigan State University.

NOTES

page 2 Beside them fluttered the mud-larks Mayhew, p. 150.

page 2 Above the river, in the streets "The pure collected is used by leather-dressers and tanners, and more especially by those engaged in the manufacture of morocco and kid leather from the skins of old and young goats, of which skins great numbers are imported, and of the roans and lambskins which are the sham morocco and kids of the 'slop' leather trade, and are used by the better class of shoemakers, bookbinders, and glovers, for the inferior requirements of their business. Pure is also used by tanners, as is pigeon's dung, for the tanning of the thinner kinds of leather, such as calf-skins, for which purpose it is placed in pits with an admixture of lime and bark. In the manufacture of moroccos and roans the pure is rubbed by the hands of the workman into the skin he is dressing. This is done to 'purify' the leather, I was told by an intelligent leatherdresser, and from that term the word 'pure' has originated. The dung has astringent as well as highly alkaline, or, to use the expression of my informant, 'scouring,' qualities. When the pure has been rubbed into the flesh and grain of the skin (the 'flesh' being originally the interior, and the 'grain' the exterior part of the cuticle), and the skin, thus purified, has been hung up to be dried, the dung removes, as it were, all such moisture as, if allowed to remain, would tend to make the leather unsound or imperfectly dressed." Mayhew, p. 143.

page 2 "What world does a dead man belong to?" Dickens 1997, p. 7.

page 3 "It usually takes the bone-picker" Mayhew, p. 139.

page 4 "the most disagreeable in the whole range of manufacture" Mayhew, p. 143.

page 5 "The removal of the refuse of a large town" Mayhew, p. 159. "Now the removal of the refuse of London is no slight task, consisting, as it does, of the cleansing of 1,750 miles of streets and roads; of collecting the dust from 300,000 dust-bins; of emptying (according to the returns of the Board of Health) the same number of cesspools, and sweeping near upon 3,000,000 chimneys." Mayhew, p. 162.

page 5 the Colosseum served as a de facto quarry Rathje and Murphy, p. 192.

page 7 But if the bacteria disappeared overnight "In fact, so significant are bacteria and their evolution that the fundamental division in forms of life on Earth is not that between plants and animals, as is commonly assumed, but between prokaryotes—organisms composed of cells with no nucleus, that is, bacteria—and eukaryotes—all the other life forms. In their first two billion years on Earth, prokaryotes continuously transformed the Earth's surface and atmosphere. They invented all of life's essential, miniaturized chemical systems—achievements that so far humanity has not approached. This ancient high biotechnology led to the development of fermentation, photosynthesis, oxygen breathing, and the removal of nitrogen gas from the air. It also led to worldwide crises of starvation, pollution, and extinction long before the dawn of larger forms of life." Margulis, p. 28.

page 8 No extended description of London *Punch* (27, September 2, 1854, p. 102) even captured the stench of the metropolis in verse:

> In every street is a yawning sewer;
> In every court is a gutter impure;
> The river runs stinking, and all its brink
> Is a fringe of every delectable stink:
> Bone-boilers and gas-workers and gut-makers there
> Are poisoning earth and polluting air.

But touch them who dares; prevent them who can;
What is the Health to the Wealth of man?

page 9 drowned in human shit Halliday 1999, p. 119.

page 10 "I found whole areas of the cellars" Halliday 1999, p. 40.

page 10 "a heap of dung" Picard, p. 60.

page 10 "We then journeyed on to London-street" Mayhew, London *Morning Chronicle,* September 24, 1849.

page 11 The visitors no doubt marveled Halliday 1999, p. 42.

page 13 "The corpses [of the poor]" Engels, p. 55.

page 13 "up to my knees in human flesh" Picard, p. 297.

page 14 "a hemmed-in churchyard" Dickens 1996, p. 165.

page 14 "There is no document of civilization" Benjamin, p. 256.

page 16 Eventually, the city's inexorable drive Summers, pp. 15–17.

page 17 Another Blake brother opened a bakery Summers, p. 121.

page 17 "In that quarter of London" Charles Dickens, *Nicholas Nickleby* (London: Penguin, 1999), pp. 162–63.

page 18 "[The flat] has two rooms" Quoted in Summers, p. 91.

page 21 Sometime in the late 1840s Vinten-Johansen et al., p. 283.

page 22 Plagues and political unrest The radical democrat James Kay-Shuttleworth described cholera as an opportunity to explore "the abodes of poverty . . . the close alleys, the crowded courts, the over-peopled habitations of wretchedness, where pauperism and disease congregate round the source of social discontent and political disorder in the centre of our large towns, and behold with alarm, in the hotbed of pestilence, ills that fester in secret, at the very heart of society." Quoted in Vinten-Johansen et al., p. 170.

page 26 "Mind you, the man" Rawnsley, p. 4.

page 26 "One does not realize" Rawnsley, p. 32.

pages 27–28 In some cases, cows were lifted Picard, p. 2.

page 28 defining the region that the "gentleman" Rawnsley, p. 34.

page 28 forced to perform arduous labor Workhouses had existed in one form or another for centuries, but the Poor Law Amendment Act of 1834 had greatly increased their number, and the severity of the "punishment" they dealt out to the pauper classes of the day. "Under the new Act, the threat of the Union workhouse was intended . . . as a deterrent to the able-bodied pauper. This was a principle enshrined in the revival of the 'workhouse test'—poor relief would only be granted to those desperate enough to face entering the repugnant conditions of the workhouse. If an able-bodied man entered the workhouse, his whole family had to enter with him. Life inside the workhouse was . . . to be as off-putting as possible. Men, women, children, the infirm, and the able-bodied were housed separately and given very basic and monotonous food such as gruel, or bread and cheese. All inmates had to wear the rough workhouse uniform and sleep in communal dormitories. Supervised baths were given once a week. The able-bodied were given hard work such as stone-breaking or picking apart old ropes. . . . The elderly and infirm sat around in the day-rooms or sick-wards with little opportunity for visitors. Parents were . . . allowed limited contact with their children—perhaps for an hour or so a week on Sunday afternoon." See http://www.workhouses.org.uk/.

page 29 "the noisy and the eager" Charles Dickens, *Little Dorrit* (London: Wordsworth, 1996), p. 778.

page 33 "burst forth . . . with extraordinary malignity" London *Times,* September 12, 1849, p. 2.

page 34 The epidemic of 1848–1849 Koch, p. 42.

pages 34–35 "While the mechanism of life" London *Times,* September 13, 1849, p. 6.

page 35 "countenance quite shrunk" Shephard, p. 158.

page 36 With the exception of a few unusual compounds "Louis Pasteur, who proved the microbial origin of such devastating diseases as foot and mouth disease, plague, and wine rot, set the tone of the relationship from the start. The context of the encounter between intellect and bacteria defined medicine as a battleground: bacteria were seen as 'germs' to be destroyed. Only today have we begun to appreciate the fact that bacteria are normal and necessary for the human body and that health is not so much a matter of destroying microorganisms as it is of restoring appropriate microbial communities." Margulis, p. 95.

page 37 A glass of water could easily contain Most of the information on the size, visibility, and replication rate of *Vibrio cholerae* comes from an interview with Harvard's John Mekalanos. The Centers for Disease Control have an excellent overview of cholera, available online at http:www.cdc.gov/ncidod/dbmd/diseaseinfo/cholera_g.htm.

page 38 "Those animal species that fully adapted" Margulis, p. 183.

page 41 "We are living at a period" Quoted in Picard, p. 215. While the Great Exhibition is more famous than the Broad Street epidemic, in a strange sense the two events have a comparable, if inverted, symbolic value: the Exhibition marking the emergence of a truly global culture, with all the dynamism and diversity that suggests, and Broad Street marking the emergence of a metropolitan culture, with all the promise and peril that offered. The twentieth century would ultimately be the story of increasingly large cities increasingly connected to one another; the Great Exhibition and Broad Street each in their separate ways helped make that a reality.

page 43 "All the world's bacteria essentially" Margulis, p. 30.

page 45 Thomas Latta, hit upon Shephard, p. 158.

page 46 "among the first to recognize" Standage, p. 234. "The Elixir of Life sold by a Dr. Kidd, for example, claimed to cure 'every known ailment. . . . The lame have thrown away crutches and walked after two or three trials of the remedy. . . . Rheumatism, neuralgia, stomach, heart, liver, kidney, blood and skin diseases disappear as by magic.' The newspapers that printed such advertisements did not ask any questions. They welcomed the advertising revenues, which enabled the newspaper industry to expand enormously. . . . The makers of St. Jacob's Oil, which was said to remedy 'sore muscles,' spent five hundred thousand dollars on advertising in 1881, and some advertisers were spending more than one million dollars a year by 1895."

page 47 "FEVER and CHOLERA" London *Morning Chronicle,* September 7, 1854.

page 47 "Sir—I have observed" London *Morning Chronicle,* August 25, 1854.

page 48 "Will you . . . kindly allow" London *Times,* August 18, 1854, p. 9.

page 49 "Sir—Induced by" London *Times,* September 21, 1854, p. 7.

page 50 "It really is nauseating" *Punch,* 27 (September 2, 1854), p. 86.

page 51 "Having at length emerged" London *Morning Chronicle,* September 1, 1854, p. 4.

page 52 Overnight, Henry Whitehead's sociable rounds Henry Whitehead's experiences and thoughts presented here are drawn almost entirely from four overlapping accounts of the epidemic authored by Whitehead himself: *The Cholera in Berwick Street,* his original pamphlet published shortly after the outbreak's conclusion; his official report for the Cholera Inquiry Committee, published the following year; an essay recalling the outbreak published in *Macmillan's Magazine* in 1865; and the transcript of an astonishingly long speech delivered at a farewell dinner on the eve of his departing London in 1873, published in H. D. Rawnsley's biography in 1898.

page 54 All but one would perish Whitehead 1854, p. 5.

page 58 But one Soho resident The details of John Snow's investigation of the Broad Street outbreak are drawn primarily from his account of the outbreak and its aftermath, in his report published in the Cholera Inquiry Committee report of 1855, and in his revised monograph, *On the Mode and Communication of Cholera.*

page 59 He would largely avoid meat Details on Snow's life up to his cholera investigations are drawn from four primary sources: Richardson's hagiographic "Life of John Snow," published shortly after Snow's death; David Shephard's biography *John Snow: Anaesthetist to a Queen and Epidemiologist to a Nation*; the superb *Cholera, Chloroform, and the Science of Medicine*; and Ralph Frerichs' invaluable John Snow Web archive hosted by UCLA's School of Public Health.

page 60 A university degree opened "With a consulting practice and beds in one of the London teaching hospitals for his patients, a man of the right character and background could achieve fame of a sort treating high society. The lure of beds in a private hospital or a nursing home where they could treat wealthy feepaying patients tempted not a few physicians. For them a university degree—the M.A. as well as the M.D., perhaps, especially from Oxford or Cambridge—was important not so much for its academic kudos as for its social cachet, because if one wished to practise in fashionable circles it was as important to be seen as a gentleman as much as a well-trained doctor. A knowledge of Latin and Greek was as much an entree to this type of practice as a knowledge of medicine itself." Shephard, p. 21.

page 61 His first published paper "The arsenic candles investigations show Snow as a collateral scientist in keeping with the new scientific approaches to medicine that were part and parcel of his training. His approach to these investigations also reveals a model that would recur in his anesthesia and cholera research. At an early stage in his career he demonstrated an ability to set up a series of experiments that traced an agent as it circulated in a medical school dissection room, in rooms where arsenic candles were burned, and in the bodies of everyone who entered them. That is, he was already concerned with chemical analysis, employing animal experimentation, and asking questions about what he would later term modes of communication—the pathways by which a specific poison was introduced into a community and where and how it lodged in the body." Vinten-Johansen et al., p. 73.

page 61 "Mr. Snow might better employ himself" "[*Lancet* editor] Wakley's statement can be read as a snub: Snow was an upstart trying to make a name for himself by finding fault with his elders. It can also be read as the reaction of a prickly editor who thought Snow was criticizing him for including flawed articles in his journal, and it can be read as a gentle, if ham-fisted, warning by a senior colleague that Snow should temper himself at so early a stage of his career. Whatever Wakley's intent, his comment was patently unfair to Snow. His first letter to the editor had detailed arsenic experiments, and the *Lancet* had reported on Westminster Society meetings at which Snow had read several papers on his research activities. He appears to have taken offense, for he found a friendlier reception in [the *London Medical Gazette*]." Vinten-Johansen et al., p. 89.

page 63 "When the dreadful steel was plunged" "Elective surgery was performed very infrequently prior to the advent of effective anesthesia. From 1821 to 1846, the annual reports of Massachusetts General Hospital recorded 333 surgeries, representing barely more than one case per month. Surgery was a last and desperate resort. Reminiscing in 1897 about preanesthesia surgery, one elderly Boston physician could only compare it to the Spanish Inquisition. He recalled 'yells and screams, most horrible in my memory now, after an interval of so many years. . . . In one of these operations, performed by the hospital's senior surgeon, John Collins Warren, M.D., the cancerous end of a young man's tongue was cut off by a sudden, swift stroke of the knife, and then a red-hot iron was placed on the wound to cauterize it. Driven frantic by the pain and the sizzle of searing flesh inside his mouth, the young man escaped his restraints in an explosive effort and had to be pursued until the cauterization was complete, with his lower lip burned in the process." Sullivan 1996.

page 65 He reaches for his pen Snow's first biographer, Richardson, reported that Snow had investigated the following agents: "carnoic, acide, carbonic oxide, cyanogen, hydrocanic acid, Dutch liquid, ammonia, nitrogen, amylovinic ether, puff-ball smoke, allyle, cyanide of ethyle, chloride of amyle, a carbo-hydrogen coming over with amylene." He went on to note: "If the agent seemed to promise favourably from these inquiries, he commenced to try it on man; and the first man was invariably his own self." Richardson, p. xxviii.

page 66 "Thursday 7 April" Snow and Ellis, p. 271.

page 67 "The Consilience of Inductions," Whewell wrote Quoted in Wilson, p. 8.

page 68 His mind tripped happily Vinten-Johansen et al. make this point with typical eloquence: "Snow was a systems-network type of reasoner. He seldom dealt with linear chains of cause and effect but rather with interacting networks of causes and effects. He viewed the human organism, and the world it inhabits, as a complex system of interacting variables, any one of which, isolated temporarily for careful study, might provide a useful clue to the clinical-scientific problem—but only when seen in its proper context, and only when the variable, having once been isolated for study, was then put back into its place in the system and restudied in its natural environment. Vinten-Johansen et al., p. 95.

page 69 "We can only suppose the existence" "History of the Rise, Progress, Ravages etc. of the Blue Cholera of India," *Lancet,* 1831, pp. 241–84.

page 70 By the time the epidemic wound down Nearly all the details of cholera outbreaks—and Snow's investigations of them—leading up to the Broad Street affair are drawn from Snow's own accounts, published in the various editions of "On the Mode and Communication of Cholera."

page 74 it didn't include the false leads J. M. Eyler, "The Changing Assessments of John Snow's and William Farr's Cholera Studies," *Sozial- und Präventivmedizin* 46 (2001), pp. 225–32.

page 75 "The *experimentum crucis* would be" *London Medical Gazette* 9 (1849), p. 466.

page 83 The papers of the day were filled In the Central London area, postal deliveries could sometimes take only an hour to reach their destination. Each residence could expect twelve regular deliveries on a weekday. Picard, p. 68.

page 83 "It is said that Friday night" *Observer,* September 3, 1854, p. 5.

page 84 The 1842 study found Picard, p. 180.

page 85 "Jo lives—that is to say" Dickens 1996, p. 475.

page 86 "The roads, in all directions" Quoted in Rosenberg 1987, p. 28.

page 88 "The infinite number of Fires" Quoted in Porter, p. 162.

page 89 "the houses will become too numerous" Porter, p. 164.

page 91 The unplanned . . . engineering of ant colonies For more on the connection between the bottom-up organization and intelligence of ant colonies and the collective development of cities, see my 2001 book *Emergence.* The extended Wordsworth quote reads: "Rise up, thou monstrous anthill on the plain / Of a too busy world! Before me flow / Thou endless stream of men and moving things! / Thy every-day appearance, as it strikes— / With wonder heightened, or sublimed by awe— / On strangers, of all ages; the quick dance / Of colours, lights, and forms . . ."

page 91 "monster city . . . stretched not only" Quoted in Porter, p. 186.

page 93 The Londoner enjoying a cup of tea For a thorough—and thoroughly entertaining—overview of the sociohistorical impact of tea (along with other beverages) see Standage's *A History of the World in Six Glasses.*

page 93 A collection of water molecules Iberall 1987, pp. 531–33.

page 94 In a sense, the Industrial Revolution "If the steam-powered factory, producing for the world market, was the first factor that tended to increase the area of urban congestion, the new railroad transportation system, after 1830, greatly abetted it. Power was concentrated on the coal fields. Where coal could be mined or obtained by cheap means of transportation, industry could produce regularly throughout the year without stoppages through seasonal failure of power. In a business system based upon time-contracts and time-payments, this regularity was highly important. Coal and iron thus exercised a gravitational pull on many subsidiary and accessory industries: first by means of the canal, and after 1830, through the new railroads. A direct connection with the mining areas was a prime condition of urban concentration: until our own day the chief commodity carried by the railroads was coal for heat and power." Mumford, p. 457.

page 95 One mechanic who provided Picard, p. 82.

page 95 Largely freed from waterborne disease Standage, p. 201.

page 99 John Snow would go to his grave A comprehensive overview of the discovery of the cholera bacterium, including a biographical sketch of Pacini himself, is available online at the UCLA John Snow archive at http:www.ph.ucla.edu/EPI/snow/firstdiscoveredcholera.html.

page 101 By the mid-1840s, his reports "He approached the Presidents of the Royal Colleges of Physicians and Surgeons and the Master of the Society of Apothecaries and persuaded them to write to their members throughout the kingdom, urging them 'to give, in every instance which may fall under our care, an authentic name of the fatal disease,' to be recorded in the local register books from which Farr compiled his statistics. At the same time, Farr compiled a 'statistical nosology,' which listed and defined 27 fatal disease categories to be used by local registrars when recording causes of death. Thus dysentery ('bloody flux') was distinguished from diarrhea ('looseness, purging, bowel complaint'). Farr also gave the 'synonymes' (sic) and 'provincial terms' by which the diseases might be known locally. Letters were drafted in the name of the Registrar-General setting the qualifications which were necessary for local registrars, and instructions were also issued to ships' captains concerning their responsibilities." Halliday 2000, p. 223.

page 102 "To measure the effects of good or bad" Quoted in Vinten-Johansen et al., p. 160. The authors offer this instructive commentary on the phrase itself: "Farr's usage of the same Baconian term that Snow had employed in his first publication indicates the importance of the hypotheticodeductive method to some medical men of this generation. In the laboratory one can conduct a 'crucial experiment' in which two samples are treated in identical fashion except for the factor in dispute. The results of the experiment then tell one with certainty whether the underlying theory is correct, but London was not a laboratory."

page 103 To digest large quantities of it Ridley, p. 192.

page 104 One provides the fizz, the other the buzz Margulis, p. 75.

page 105 S&V chose to delay its move In many ways, Snow's "grand experiment" with the metropolitan water supply stands as a more impressive—and, arguably, more convincing—example of medical sleuthing than the Broad Street case. For a detailed account, see Vinten-Johansen et al., pp. 254–82.

page 106 "The experiment . . . was on the grandest" Snow, 1855a, p. 75.

page 109 "In Broad-Street, on Monday evening" *Observer,* September 3, 1854, p. 5.

page 112 "The Guardians are acting" London *Times,* September 6, 1854, p. 5.

page 114 This is the great irony of Chadwick's life For more on the life of Chadwick, see Finer.

page 114 "All smell is . . . disease" Quoted in Halliday 1999, p. 127.

page 115 One in twenty had human waste Halliday 1999, p. 133.

page 116 "According to the average of the returns" Mayhew could also wax philosophical on these issues, in language that was strikingly ahead of its time: "Now, in Nature everything moves in a circle—perpetually changing, and yet ever returning to the point whence it started. Our bodies are continually decomposing and recomposing—indeed, the very process of breathing is but one of decomposition. As animals live on vegetables, even so is the refuse of the animal the vegetable's food. The carbonic acid which comes from our lungs, and which is poison for us to inhale, is not only the vital air of plants, but positively their nutriment. With the same wondrous economy that marks all creation, it has been ordained that what is unfitted for the support of the superior organisms, is of all substances the best adapted to give strength and vigour to the inferior. That which we excrete as pollution to our system, they secrete as nourishment to theirs. Plants are not only Nature's scavengers but Nature's purifiers. They remove the filth from the earth, as well as disinfect the atmosphere, and fit it to be breathed by a higher order of beings. Without the vegetable creation the animal could neither have been nor be. Plants not only fitted the earth originally for the residence of man and the brute, but to this day they continue to render it habitable to us. For this end their nature has been made the very antithesis to ours. The process by which we live is the process by which they are destroyed. That which supports respiration in us produces putrefaction in them. What our lungs throw off, their lungs absorb—what our bodies reject, their roots imbibe. . . . In every well-regulated State, therefore, an effective and rapid means for carrying off the ordure of the people to a locality where it may be fruitful instead of destructive, becomes a most important consideration. Both the health and the wealth of the nation depend upon it. If to make two blades of wheat grow where one grew before is to confer a benefit on the world, surely to remove that which will enable us at once to do this, and to purify the very air which we breathe, as well as the water which we drink, must be a still greater boon to society. It is, in fact, to give the community not only a double amount of food, but a double amount of health to enjoy it. We are now beginning to understand this. Up to the present time we have only thought of removing our refuse—the idea of using it never entered our minds. It was not until science taught us the dependence of one order of creation upon another, that we began to see that what appeared worse than worthless to us was Nature's capital—wealth set aside for future production." Mayhew, p. 160.

page 116 He also entertained an aquatic version Another visionary named William Hope thought that these new sewage farms might attract visitors as a kind of excrement-themed spa: "London beauties might come out to recruit their wasted energies at the close of the season, and . . . would perhaps at times listen to a lecture on agriculture from the farmer himself, while drinking his cream and luxuriating in the health-restoring breeze." Halliday 1999, p. 133.

page 118 "[Any] Dwelling House or Building" Nuisances Act, September 4, 1848, p. 1.

page 119 the sewers themselves began to clog Halliday 1999, pp. 30–34.

page 120 "The Thames is now made" Halliday 1999, p. 35.

page 121 "On entering the precincts" "A Visit to the Cholera Districts of Bermondsey," London *Morning Chronicle,* September 24, 1849.

page 122 "How is the cholera generated?" London *Times,* September 13, 1854, p. 6.

pages 122–23 "telluric theory" . . . "failed to include all the observed phenomena" London *Times,* September 13, 1849, p. 6.

page 123 "The very first canon of nursing" Florence Nightingale, *Notes on Nursing* (New York: Dover, 1969), p. 12.

page 124 "If the tell-tale air test" Nightingale, p. 17.

page 125 "It might be supposed" Mayhew, p. 152.

page 127 "Whoever wishes to investigate" Hippocrates, p. 4.

page 127 "the atmosphere, all over the world" Whitehead 1854, p. 13.

page 128 The brain scans in the 2003 study Royet et al., pp. 724–26.

page 132 For every sewer-hunter living happily Tom Koch offers a precise and articulate survey of some of the statistical and cartographic studies offered in defense of the miasma theory during this period, including work on elevation authored by Farr. In most cases, Koch observes, the studies were thorough and internally consistent, even if they were ultimately supporting an incorrect hypothesis. "While the miasmatic, contagionist conclusion was wrong, the inverse relationship that was used to argue it was accurate. That Acland and Farr missed the meaning of the relation is a fault neither of the researchers nor the mapping they did. In contention were different theories of disease, different perceptions of the city, and different assumptions about the data required for a disease study. One cannot blame a scientist for being limited by the science and knowledge of his time." Koch, p. 126.

page 133 "The probability of an outburst" Quoted in Vinten-Johansen et al., p. 174.

page 142 Snow noticed another telling absence There is some ambiguity about the timing of these investigations in the historical record. Snow's investigation of Broad Street unfolded in two primary phases: a rapid survey of the neighborhood as the outbreak was still raging, and then a longer study that commenced a few weeks after the outbreak subsided, based partly on second-hand accounts from other surgeons and physicians in the area. Snow may in fact have uncovered information about the brewery and the workhouse in his later investigation, though the prominence of both operations, in terms of number of employees and proximity to the pump, makes it more likely that Snow paid them both a visit during the outbreak itself. In his published account Snow merely reports: "There is a Brewery in Broad Street, near to the pump, and on perceiving that no brewer's men were registered as having died of cholera, I called on Mr. Huggins, the proprietor." This appears several paragraphs after his description of requesting the *Weekly Returns* from the Registrar-General's Office shortly after September 2.

page 145 Snow was naturally inclined to view the theory "Perhaps his research into the nature and mechanisms of anesthesia by inhaled gases made him certain that gaseous vapors alone, whether general or local, could not cause specific epidemic diseases, as miasmatic theory posited. Moreover, his investigation of arsenical candles had suggested that when a body inhaled a specific poison, it showed the specific effects of that poison, not the generalized fevers typically claimed for miasmatic and local effluvial poisoning. Contrary to the older generation of medical men who dismissed the law of the diffusion of gases as armchair theorizing, Snow's training and daily experience administering anesthesia made him believe that careful attention to the chemistry and physics of gases could have practical benefits. It was precisely that which permitted him to use otherwise dangerous medicinal agents with safety and with exact application to the peculiar needs of each patient and each surgical operation." Vinten-Johansen et al., p. 202.

page 145 "I have arrived at the conclusion" Lilienfeld, p. 5.

page 146 For Snow . . . an obvious etiology "A consideration of the pathology of cholera is capable of indicating to us the manner in which the disease is communicated. If it were ushered in by fever, or any other general constitutional disorder, then we should be furnished with no clue to the way in which the morbid poison enters the system; whether, for instance, by the alimentary canal, by the lungs, or in some other manner, but should be left to determine this point by circumstances unconnected with the pathology of the disease. But from all that I have been able to learn of cholera, both from my own observations and the descriptions of others, I conclude that cholera invariably commences with the affection of the alimentary canal. The disease often proceeds with so little feeling of general illness, that the patient does not consider himself in danger, or even apply for advice, till the malady is far advanced. In a few cases, indeed, there are dizziness, faintness, and a feeling of sinking, before discharges from the stomach or bowels actually take place; but there can be no doubt that these symptoms depend on the exudation from the mucous membrane, which is soon afterwards copiously evacuated." Snow 1855a, pp. 6–9.

page 149 He delivered chloroform to two patients Snow's casebooks report the full range of his professional activity for the week: "Saturday 2 Administered Chloroform at Mr Duffins to a little girl three years old from the neighbourhood of Blackheath whilst Mr D. performed amputation of the great toe together with its metatarsal bone. Monday 4 Administered Chloroform at Mr Cartwright's to a lady whilst he extracted two [?] teeth. Wednesday 6 Administered Chloroform to Mr Jenner, Linen draper, Edgware Road whilst Mr Salmon operated by ligatures on some haemorrhoids. The patient was extremely blanched from loss of blood from the past and had a bounding haemorrhagic pulse. No faintness or depression from the chloroform. Administered Chloroform at 16 Hanover Square whilst Mr A. Rogers extracted 2 teeth. Thursday 7 Administered Chloroform to a gentleman on King Street Covent garden patient of Mr Edwards whilst Mr Partridge operated for haemorrhoid. No sickness &c. Friday 8 Administered Chloroform at 46 Wigmore Street whilst Mr Salmon operated for fistula in ano. No sickness." Snow and Ellis, pp. 342–43.

page 152 But the most likely scenario I am grateful to Harvard's John Mekalanos for suggesting this scenario.

page 154 "No one but those who knew him" Richardson, p. xix.

page 154 St. Bartholomew's Hospital had received *Lancet,* September 16, 1854, p. 244.

page 160 And so . . . the Board voted Snow's own description of the exchange is taciturn: "I had an interview with the Board of Guardians of St. James's parish, on the evening of Thursday, 7th September, and represented the above circumstances to them. In consequence of what I said, the handle of the pump was removed on the following day." This last sentence is now memorialized on a pin worn by members of the John Snow Society. Snow 1855a.

page 160 "Owing to the favourable change in the weather" *Globe,* September 8, 1854, p. 3.

page 161 "We regret to announce" *Globe,* September 9, 1854, p. 3.

page 162 These were real achievements Richardson probably did more than anyone to build the story that the pump handle's removal had single-handedly brought the outbreak to an end. "The pump handle was removed," he triumphantly announced, "and the plague was stayed." The popular version of the Broad Street story conventionally follows this appealing narrative line. Snow identifies the perpetrator, and brings its reign of terror to an immediate end. In my research, nearly half of the shorthand accounts of the outbreak tell the story along these lines.

Snow did not demonstrate the link between the pump and the cholera by removing the handle; he demonstrated the link through statistical analysis of data accumulated via door-to-door interviews. And of course, the pump was not the neighborhood's only water source, merely the most popular. In fact, the existence of the other water sources was crucial to Snow's case. But the biggest—and most common—distortion is the notion that closing down the pump single-handedly brought the outbreak to an end. Removing the pump, in all likelihood, had little impact on the course of the outbreak. New attacks were already on the wane before Snow had the handle removed, and it's entirely possible that the water had ceased to be dangerous by the time the authorities did anything about it.

The final statistics for the Broad Street outbreak suggest that the removal of the pump handle likely played a minor role in the ultimate trajectory of the outbreak. The most dramatic decline in deaths falls between the 4th and 5th of September, while the second-most dramatic drop occurs between the 10th and the 12th. The timeline of attacks, not deaths, has a more dramatic spike at the beginning of the week, followed by a steady leveling-off. The number of new attacks reaches the statistical norm for the neighborhood only by the 12th. If you assume a twenty-four-to-forty-eight-hour incubation period between ingesting *V. cholerae* and the first onset of symptoms, it would appear that the closing of the Broad Street pump well may have extinguished what was left of the outbreak, like a fire department arriving to snuff out the last embers of a building that has already burned to the ground. The plague may well have been stayed by Snow's intervention, but it was already on its last legs. However, as we will see at the end of this chapter, there might well have been a renewed epidemic after John Lewis contracted the disease had Snow not convinced the authorities to shut down the pump.

page 163 "Structural peculiarities of the Streets" Committee for Scientific Inquiries, pp. 138–64.

page 169 "Dufour's Place . . . Five houses escaped" Whitehead 1854, p. 4.

page 170 "There were no less than 21 instances" Whitehead 1854, p. 6.

page 170 "God's ways are equal" Whitehead 1854, p. 14.

page 172 "principally on the ground" Cholera Inquiry Committee, p. v.

page 175 As much as he had resisted Whitehead described his response to Snow's theory in his 1865 memoir: "When I first heard of it, I stated to a medical friend my belief that a careful investigation would refute it, alleging as one proof of its inaccuracy the fact of several recoveries from collapse having taken place, at least in spite of, if not actually by reason of, the constant use of the Broad Street water. I added that I knew the inhabitants of Broad Street so well, and had occasion almost daily to spend so much time among them, that I should have no great difficulty in making the necessary inquiries. Accordingly I began an inquiry, which ultimately became very elaborate; at an early stage of which, however, one day meeting the same friend, and being asked by him what way I had made towards clearing the character of the pump, I was obliged to confess that my opinion on that matter was less confident than when we had last talked about it." Whitehead 1865, p. 116.

page 176 Whatever agent had caused the cholera Whitehead 1865, p. 116.

page 179 "abominations, unmolested by water" Whitehead 1865, p. 121.

page 181 "You and I may not live" Rawnsley, p. 206.

page 182 "The weight of both positive and negative" Cholera Inquiry Committee, p. 55.

page 183 "In explanation of the remarkable intensity" Committee for Scientific Inquiries, p. 51.

page 184 "That such local uncleanliness" Committee for Scientific Inquiries, p. 52.

page 185 "Atmospheric Pressure" Committee for Scientific Inquiries, p. iv.

page 186 "The water was undeniably impure" Committee for Scientific Inquiries, p. 52.

page 192 If some noxious effluvium Koch, pp. 106–8.

page 194 It was not the mapmaking technique Koch, pp. 75–101. Vinten-Johansen et al. also have a superb chapter on Snow's cartographic legacy that addresses many of these topics.

page 196 it measured how long it took Koch, p. 100.

page 198 copies of copies began appearing in textbooks The original redrawing appears in Sedgewick's public-health textbook from 1911. For a meticulous investigation of the Broad Street map's convoluted history, see Koch, pp. 129–53.

page 204 Snow responded to these papers "SIR,—I did not until to-day, read the important and interesting Address of Sir J. K. Shuttleworth, Bart., in

The Lancet of the 2nd instant. I find that he alludes in complimentary terms to my conclusions regarding the propagation of cholera, as modified by a suggestion of Drs. Theirsch and Pettenkofer, but he erroneously attributes these views, so modified, to Dr. W. Budd. . . . A few weeks after the first edition of my essay on Cholera appeared in 1849, Dr. W. Budd published a pamphlet on the subject, in which he adopted my views, and made a full and handsome acknowledgement of my priority." *Lancet,* February 16, 1856, p. 184.

page 205 "Why is it, then, that Dr. Snow" *Lancet,* June 23, 1855, p. 635.

page 205 "What a pity" Quoted in Halliday 1999, p. 82.

page 206 "DR. JOHN SNOW—This well-known physician" *Lancet,* June 26, 1858, p. 635.

page 208 "It was certainly a very troublesome job" Quoted in Halliday 1999, p. 183.

page 211 Ninety-three percent of the dead This account of the East London outbreak is drawn largely from Halliday 1999, pp. 137–43.

page 212 "The final report of the scientific committee" Parliamentary Papers, 1867–1868, vol. 37, pp. 79–82.

page 215 to reverse the flow of the Chicago River http:www.sewerhistory.org/chronos/new_amer_roots.htm.

page 216 The main road in . . . Sultaneyli Neuwirth, pp. 1–11.

page 218 A service called GeoSentinel http:www.istm.org/geosentinel/main.html.

page 221 "Towns and suburbs . . . are natural homes" Jacobs 1969, pp. 146–47. The current buzzword for this trend is "long tail" economics; instead of concentrating exclusively on big mass hits, online businesses can target the "long tail" of quirkier fare. In the old model, the economics dictated that it was always better to sell a million copies of one album. But in the digital age, it can be just as profitable to sell a hundred copies each of a thousand different albums. Urban information mapping systems offer an intriguing corollary to the long-tail theory. As technology increasingly allows us to satisfy more eclectic needs, anytime those needs require physical presence, the logic of the long tail will favor urban environments over less densely populated ones. If you're downloading the latest album from an obscure Scandinavian doo-wop group, geography doesn't matter: it's just as easy to get the bits delivered to you in the middle of Wyoming as it is in the middle of Manhattan. But if you're trying to meet up with other fans of Scandinavian doo-wop, you'll have more luck in Manhattan or London. The long tail may well lead us away from the dominance of mass hits and pop superstars toward quirkier tastes and smaller artists. But it may also lead us to bigger cities.

page 225 The public spaces and coffeehouses "'The coffee-house was the Londoner's home, and those who wished to find a gentleman commonly asked, not whether he lived in Fleet Street or Chancery Lane, but whether he frequented the Grecian or the Rainbow.' Some people frequented multiple coffeehouses, the choice of which depended on their interests. A merchant, for example, might oscillate between a financial coffeehouse and one specializing in Baltic, West Indian, or East Indian shipping. The wide-ranging interests of the English scientist Robert Hooke were reflected in his visits to around sixty London coffeehouses during the 1670s, recorded in his diary. Rumors, news, and gossip were carried between coffeehouses by their patrons, and on occasion runners would flit from one coffeehouse to another to report major events such as the outbreak of war or the death of a head of state." Standage, p. 155.

page 226 "In the Broad Street outbreak" Quoted in Rawnsley, p. 76.

page 227 "that in any profession the highest order" Rawnsley, p. 206.

page 232 Two-thirds of the women living in rural areas Statistics from "State of World Population 1996." See http://www.unfpa.org/swp/1996/.

page 233 "Virtually any service system" Toby Hemenway, "Cities, Peak Oil, and Sustainability." Published at http://www.patternliteracy.com/urban2.html.

page 234 If we're going to survive as a planet Much has been made of the staggering size of the environmental footprint of today's modern city, the area of land required to support sustainably the energy intakes of the city's population. London's environmental footprint, for instance, is practically as large as the entire United Kingdom. The sheer magnitude of such a footprint has been invoked as part of antiurban environmental arguments, but the primary objection is in fact industrialization not urbanization. However large London's footprint might be today, it would be many times larger if the city's population were scattered at suburban or exurban densities. Unless we renounce our postindustrial lifestyle altogether, cities are environmentally preferable to other, lower-density forms of living. The United Nations' *Global Environmental Outlook* describes it this way: "The relatively disproportionate urban environmental footprint is acceptable to a certain extent because, for some issues, the per capita environmental impact of cities is smaller than would be made by a similar number of people in a rural setting. Cities concentrate populations in a way that reduces land pressure and provides economies of scale and proximity of infrastructure and services. . . . Urban areas therefore hold promise for sustainable development because of their ability to support a large number of people while limiting their per capita impact on the natural environment."

page 235 "All the apparatus of surgery" Jacobs 1969, pp. 447–48.

page 238 "The most devastating damage" Owen, p. 47. Owen describes the environmental impact of his family's move from Manhattan to rural northwest Connecticut: "Yet our move was an ecological catastrophe. Our consumption of electricity went from roughly four thousand kilowatt-hours a year, toward the end of our time in New York, to almost thirty thousand kilowatt-hours in 2003—and our house doesn't even have central air-conditioning. We bought a car shortly before we moved, and another one soon after we arrived, and a third one ten years later. (If you live in the country and don't have a second car, you can't retrieve your first car from the mechanic after it's been repaired; the third car was the product of a mild midlife crisis, but soon evolved into a necessity.) My wife and I both work at home, but we manage to drive thirty thousand miles a year between us, mostly doing ordinary errands. Nearly everything we do away from our house requires a car trip. Renting a movie and later returning it, for example, consumes almost two gallons of gasoline, since the nearest Blockbuster is ten miles away and each transaction involves two round trips. When we lived in New York, heat escaping from our apartment helped to heat the apartment above ours; nowadays, many of the BTUs produced by our brand-new, extremely efficient oil-burning furnace leak through our two-hundred-year-old roof and into the dazzling star-filled winter sky above."

page 240 But we don't have that option One "third-way" solution to this problem would be to adopt the medieval system of distributed density, still visible in hill towns of northern Italy: a network of tightly packed mixed-use nodes of finite size, separated by large stretches of low-density vineyards and farms. This is not the decentralized approach of edge-of-city sprawl; the towns in the medieval system were not as dense and economically diverse as most modern city centers, but they had a ceiling on their overall growth, usually defined by the walls that outlined the town limits. A post-9/11 city could be built along similar lines: the density of traditional metropolitan space in distributed nodes limited to 50,000 to 100,000 people each, separated by expanses of low-density development: parkland, nature preserves, sports facilites, even vineyards where the climate allows. Such a model would reverse the Olmsted vision of urban greenery: rather than carve out a park in the middle of an immense city, the new model builds a space for nature on the edges of the city center—Peripheral Park, instead of Central. In medieval times, the walls protected the town population. In these theoretical settlements, the open spaces separating the nodes would keep the city safe. Imagine a city of 2 million people, built out of twenty nodes. In a worst-case scenario, a terrorist with a backpack full of smallpox might well be able to do extensive damage to a single node, perhaps killing tens of thousands in the process—not millions. The remaining nodes would be largely unaffected, not unlike the Arpanet and its now folkloric skills at routing around damage. An attack like those on the Twin Towers could still do a lot of damage, but there wouldn't be a centralized, symbolic node to target. Life in such a metropolitan complex would not feel suburban, by any means: the generative force of sidewalk culture and urban density would be preserved, possibly even enhanced.

page 243 In September 2004, health officials in Thailand "Asian Shots Are Proposed as Flu Fighter," *New York Times,* October 13, 2005.

page 246 It needs the CTX phage to switch over Mekalanos et al., pp. 241–48.

page 253 but detection is hardly a fail-safe option I described some of the latest advances in radiation detection—and speculated on how they might be employed to defend large metropolitan areas from nuclear terrorism—in the essay "Stopping Loose Nukes," published in *Wired,* November 2002.

page 254 But if the trends of asymmetric warfare continue The one thing we can do now to prevent such a dark future is to radically reduce, if not eliminate, the current stockpiles of nuclear weapons in the world. The United States alone has around 10,000 weapons in its active arsenal. This is madness in an age of asymmetric warfare, where mutually assured destruction is meaningless. (It was madness in the cold war too, but for different reasons.) If all the nuclear powers agreed to limit their stockpiles to no more than ten weapons per country—thereby reducing the total number of weapons in the world from 20,000 to less than a hundred—we would reduce by more than an order of magnitude the risk that a weapon would fall into the wrong hands. We would still retain the ability to kill 100 million people and do untold environmental damage with those ten nukes, but at least we would be making significant progress against the growing menace of proliferation. It would be an epic undertaking, yet history shows we are capable of projects on this scale, if we apply ourselves. We eliminated smallpox from the wild, after all. If we can rid the world of a microscopic virus, we can eliminate weapons the size of tractor-trailers. We hear a lot of war-on-terror rhetoric cajoling us to be realistic about the threats that face us, to confront those threats without pity or foolish idealism. That's why we have elective wars and unauthorized wiretapping: because we're realists now, or so we're told. But wherever each of us stands on the wars and the wiretaps, we need to agree that maintaining a stockpile of 10,000 nuclear weapons is the very opposite of realism. It is, in fact, an idealism of the most starry-eyed sort: the ideal that says we're better off spending billions of dollars maintaining devices that would, were they all detonated, potentially end life as we know it on planet Earth. We are, as a species, sleeping with a gun under our pillow. It may make us feel safe to know that we have all that firepower so close at hand, but someday it's going to go off.

page 255 Angola is suffering through the worst outbreak "Angola is suffering its worst outbreak of cholera in more than a decade, recording 554 deaths and 12,052 cases in just over two months, according to Doctors Without Borders. The disease has spread unusually fast, even for Africa, where cholera epidemics are common and often hard to control, said Stephan Goetghebuer, an operational coordinator for the organization. It has set up eight clinics in Angola to treat the sick and plans to open more." "Angola Is Hit by Outbreak of Cholera," *New York Times,* April 20, 2006.

BIBLIOGRAPHY

Ackroyd, Peter. *London: The Biography.* Anchor, New York: 2000.

Barry, John M. *The Great Influenza: The Epic Story of the Deadliest Plague in History.* New York: Penguin, 2005.

Benjamin, Walter. *Illuminations.* New York: Schocken, 1986.

Bingham, P., N. O. Verlander, and M. J. Cheal. "John Snow, William Farr and the 1849 Outbreak of Cholera That Affected London: A Reworking of the Data Highlights the Importance of the Water Supply." *Public Health* 118 (2004): 387–94.

Brand, Stewart. "City Planet." http://www.strategy-business.com/press/16635507/06109.

Brody, H., et al. "John Snow Revisited: Getting a Handle on the Broad Street Pump." *Pharos Alpha Omega Alpha Honor Med. Soc.* 62 (1999): 2–8.

Buechner, Jay S., Herbert Constantine, and Annie Gjelsvik. "John Snow and the Broad Street Pump: 150 Years of Epidemiology." *Medicine & Health Rhode Island* 87 (2004): 314–15.

Cadbury, Deborah. *Dreams of Iron and Steel: Seven Wonders of the Nineteenth Century, from the Building of the London Sewers to the Panama Canal.* New York: Fourth Estate, 2004.

Chadwick, Edwin. *Report on the Sanitary Condition of the Labouring Population of*

Great Britain: A Supplementary Report on the Results of a Special Inquiry into the Practice of Interment in Towns. London, 1843.

The Challenge of Slums: Global Report on Human Settlements, 2003. Sterling, VA: Earthscan, 2003.

Cholera Inquiry Committee. *Report on the Cholera Outbreak in the Parish of St. James, Westminster, during the Autumn of 1854.* London, 1855.

Committee for Scientific Inquiries. *Report of the Committee for Scientific Inquiries in Relation to the Cholera-Epidemic of 1854.* London: HMSO, 1855.

Cooper, Edmund. "Report on an Enquiry and Examination into the State of the Drainage of the Houses Situate in That Part of the Parish of St. James, Westminster . . ." September 22, 1854.

Creaton, Heather. *Victorian Diaries: The Daily Lives of Victorian Men and Women.* London: Mitchell Beazley, 2001.

De Landa, Manuel. *A Thousand Years of Nonlinear History.* New York: Zone, 1997.

Dickens, Charles. *Bleak House.* London: Penguin, 1996.

———. *Our Mutual Friend.* New York: Penguin, 1997.

Engels, Friedrich. *The Condition of the Working Class in England.* Palo Alto, CA: Stanford University Press, 1968.

Eyler, J. M., "The Changing Assessments of John Snow's and William Farr's Cholera Studies," *Sozial- und Präventivmedizin* 46 (2001), pp. 225–32.

Farr, William. "Report on the Cholera Epidemic of 1866 in England." In U.K. Parliament, Sessional Papers, 1867–1868, vol. 37.

Faruque, S. M., M. J. Albert, and J. J. Mekalanos. "Epidemiology, Genetics, and Ecology of Toxigenic *Vibrio cholerae.*" *Microbiology and Molecular Biology Reviews* 62 (1998): 1301–14.

Faruque, Shah M., et al. "Self-Limiting Nature of Seasonal Cholera Epidemics: Role of Host-Mediated Amplification of Phage." *Proceedings of the National Academy of Science U.S.A.* 102 (2005): 6119–24.

Finer, S. E. *The Life and Times of Sir Edwin Chadwick.* New York: Barnes & Noble, 1970.

Garrett, Laurie. *The Coming Plague: Newly Emerging Diseases in a World out of Balance.* New York: Farrar, Straus & Giroux, 1994.

———. *Betrayal of Trust: The Collapse of Global Health.* New York: Oxford University Press, 2001.

Gould, Stephen Jay. *Full House: The Spread of Excellence from Plato to Darwin*. New York: Harmony, 1996.

Halliday, Stephen. *The Great Stink of London: Sir Joseph Bazalgette and the Cleansing of the Victorian Metropolis.* Phoenix Mill, England: Sutton, 1999.

———. "William Farr: Campaigning Statistician." *Journal of Medical Biography* 8 (2000): 220–27.

Häse, C. C., and J. J. Mekalanos. "TcpP Protein Is a Positive Regulator of Virulence Gene Expression in *Vibrio cholerae*." *Proceedings of the National Academy of Science U.S.A.* 95 (1998): 730–34.

Hippocrates. *Hippocrates on Airs, Waters, and Places.* Translated by Emile Littré and Janus Cornarius and Johannes Antonides van der Linden and Francis Adams. London, 1881.

Hohenberg, Paul M., and Lynn Hollen Lees. *The Making of Urban Europe, 1000–1994.* Cambridge, MA: Harvard University Press, 1995.

Iberall, Arthur S. "A Physics for Studies of Civilization." *Self-Organizing Systems: The Emergence of Order,* ed. F. Eugene Yates. New York and London: Plenum Press, 1987.

Jacobs, Jane. *The Economy of Cities.* New York: Random House, 1969.

———. *The Death and Life of Great American Cities.* New York: Vintage, 1992.

———. *The Nature of Economies.* New York: Modern Library, 2000.

Kelly, John. *The Great Mortality: An Intimate History of the Black Death, the Most Devastating Plague of All Time.* New York: HarperCollins, 2005.

Koch, Tom. *Cartographies of Disease: Maps, Mapping, and Medicine.* Redlands, CA: ESRI Press, 2005.

Kostof, Spiro. *The City Shaped: Urban Patterns and Meanings Through History.* Boston: Little, Brown, 1991.

Lilienfeld, A. M., and D. E. Lilienfeld. "John Snow, the Broad Street Pump and Modern Epidemiology." *International Journal of Epidemiology,* 1984.

Lilienfeld, D. E. "John Snow: The First Hired Gun?" *American Journal of Epidemiology* 152 (2000): 4–9.

McLeod, K. S. "Our Sense of Snow: The Myth of John Snow in Medical Geography." *Social Science in Medicine* 50 (2000): 923–35.

McNeill, William Hardy. *Plagues and Peoples.* New York: Anchor Press, 1976.

Marcus, Steven. *Engels, Manchester, and the Working Class.* New York: Norton, 1985.

Margulis, Lynn, with Dorion Sagan. *Microcosmos: Four Billion Years of Evolution from Our Microbial Ancestors.* Berkeley: University of California Press, 1997.

Mayhew, Henry. *London Labour and the London Poor.* New York: Penguin, 1985.

Mekalanos, J. J., E. J. Rubin, and M. K. Waldor. "Cholera: Molecular Basis for Emergence and Pathogenesis." *FEMS Immunol. Med. Microbiol.* 18 (1997): 241–48.

Mumford, Lewis. *The City in History: Its Origins, Its Transformations and Its Prospects.* New York and London: Harcourt Brace Jovanovich, 1961.

Neuwirth, Robert. *Shadow Cities: A Billion Squatters, a New Urban World.* New York: Routledge, 2005.

Nightingale, Florence. *Notes on Nursing: What It Is, and What It Is Not.* Philadelphia: Lippincott, 1992.

Owen, David. "Green Manhattan." *The New Yorker,* October 18, 2004.

Paneth, Nigel. "Assessing the Contributions of John Snow to Epidemiology: 150 Years After Removal of the Broad Street Pump Handle." *Epidemiology* 15 (2004): 514–16.

Picard, Liza. *Victorian London: The Life of a City, 1840–1870.* New York: St. Martin's, 2006.

Porter, Roy. *London: A Social History.* Cambridge: Harvard University Press, 1995.

Rathje, William L., and Cullen Murphy. *Rubbish! The Archaeology of Garbage.* Tucson: University of Arizona Press, 2001.

Rawnsley, Hardwicke D. *Henry Whitehead. 1825–1896: A Memorial Sketch.* Glasgow, 1898.

Richardson, Benjamin W. "The Life of John Snow." In John Snow, *On Chloroform and Other Anaesthetics,* ed. B. W. Richardson. London, 1858.

Ridley, Matt. *Genome: The Autobiography of a Species in 23 Chapters.* New York: HarperCollins, 1999.

Rogers, Richard. *Cities for a Small Planet.* Boulder, CO: Westview, 1998.

Rosenberg, Charles E. *The Cholera Years: The United States in 1832, 1849, and 1866.* Chicago: University of Chicago Press, 1987.

———. *Explaining Epidemics and Other Studies in the History of Medicine.* New York: Cambridge University Press, 1992.

Royet, Jean-P., et al. "fMRI of Emotional Responses to Odors: Influence of Hedonic Valence and Judgment, Handedness, and Gender." *Neuroimage* 20 (2003): 713–28.

Schonfeld, Erick. "Segway Creator Unveils His Next Act." *Business 2.0,* February 16, 2006.

Sedgwick, W. T. *Principles of Sanitary Science and the Public Health with Special Reference to the Causation and Prevention of Infectious Diseases.* New York, 1902.

Shephard, David A. E. *John Snow: Anaesthetist to a Queen and Epidemiologist to a Nation: A Biography.* Cornwall, Prince Edward Island: York Point, 1995.

Smith, George Davey. "Commentary: Behind the Broad Street Pump: Aetiology, Epidemiology and Prevention of Cholera in Mid-19th Century Britain." *International Journal of Epidemiology* 31 (2002): 920–32.

Snow, John. "The Principles on Which the Treatment of Cholera Should Be Based." *Medical Times and Gazette* 8 (1854a): 180–82.

———. "Communication of Cholera by Thames Water." *Medical Times and Gazette* 9 (1854b): 247–48.

———. "The Cholera Near Golden-square, and at Deptford." *Medical Times and Gazette* 9 (1854c): 321–22.

———. "On the Communication of Cholera by Impure Thames Water." *Medical Times and Gazette* 9 (1854d): 365–66.

———. *On the Mode of Communication of Cholera.* 2nd ed. London: Churchill; 1855a.

———. "Further Remarks on the Mode of Communication of Cholera; Including Some Comments on the Recent Reports on Cholera by the General Board of Health." *Medical Times and Gazette* 11 (1855b): 31–35, 84–88.

———. "On the Supposed Influence of Offensive Trades on Mortality." *Lancet* 2 (1856): 95–97.

———. "On Continuous Molecular Changes, More Particularly in Their Relation to Epidemic Diseases." London: Churchill, 1853. In *Snow on Cholera,* ed. Wade Hampton Frost. New York: Hafner, 1965.

Snow, John, and Richard H. Ellis. *The Case Books of Dr. John Snow.* London: Wellcome Institute for the History of Medicine, 1994.

Snow, John, Wade Hampton Frost, and Benjamin Ward Richardson. *Snow on Cholera: Being a Reprint of Two Papers.* New York: The Commonwealth Fund, 1965.

Specter, Michael. "Nature's Bioterrorist." *The New Yorker,* February 28, 2005: 50–62.

Standage, Tom. *A History of the World in Six Glasses.* New York: Holtzbrinck, 2005.

Stanwell-Smith, R. "The Making of an Epidemiologist." *Communicable Disease and Public Health,* 2002: 269–70.

Sullivan, John. "Surgery Before Anesthesia." *ASA Newsletter* 60.

Summers, Judith. *Soho: A History of London's Most Colourful Neighbourhood.* London: Bloomsbury, 1989.

Tufte, Edward R. *The Visual Display of Quantitative Information.* Cheshire, CT: Graphics Press, 1983.

———. *Envisioning Information.* Cheshire, CT: Graphics Press, 1990.

———. *Visual Explanations: Images and Quantities, Evidence and Narrative.* Cheshire, CT: Graphics Press, 1997.

United Kingdom General Board of Health. "Report of the Committee for Scientific Inquiries in Relation to the Cholera-Epidemic of 1854." London: HMSO, 1855.

Vandenbroucke, J. P. "Snow and the Broad Street Pump: A Rediscovery." *Lancet,* November 11, 2000, pp. 64–68.

Vandenbroucke, J. P., H. M. Eelkman Rooda, and H. Beukers. "Who Made John Snow a Hero?" *American Journal of Epidemiology* 133, no. 10 (1991): 967–73.

Vinten-Johansen, Peter, et al. *Cholera, Chloroform, and the Science of Medicine: A Life of John Snow.* New York: Oxford University Press, 2003.

White, G. L. "Epidemiologic Adventure: The Broad Street Pump." *South. Med. J.* 92 (1999): 961–62.

Whitehead, Henry. *The Cholera in Berwick Street,* 2nd ed. London: Hope & Co., 1854.

———. "The Broad Street Pump: An Episode in the Cholera Epidemic of 1854." *Macmillan's Magazine,* 1865: 113–22.

———. "The Influence of Impure Water on the Spread of Cholera." *Macmillan's Magazine,* 1866: 182–90.

Williams, Raymond. *The Country and the City.* New York: Oxford University Press, 1973.

Zimmer, Carl. *Parasite Rex: Inside the Bizarre World of Nature's Most Dangerous Creatures.* New York: Free Press, 2000.

Zinsser, Hans. *Rats, Lice, and History.* New York: Black Dog & Leventhal, 1996 (orig. pub. 1934).

INDEX

Italicized page numbers indicate illustrations.

INDEX

INDEX

INDEX

INDEX

INDEX

INDEX

INDEX

ILLUSTRATION CREDITS

Page xviii: Courtesy *Illustrated London News*.

Page 24: Courtesy General Research Division, The New York Public Library, Astor, Lenox, and Tilden Foundations.

Page 56: Courtesy Ralph R. Frerichs, UCLA Department of Epidemiology, School of Public Health, www.ph.ucla.edu/epi/snow.html.

Page 110: Courtesy National Museum of Photography, Film, & Television/ Science & Society Picture Library.

Page 138: Courtesy National Library of Medicine and Light, Inc.

Page 190: Courtesy Ralph R. Frerichs, UCLA Department of Epidemiology, School of Public Health, www.ph.ucla.edu/epi/snow.html.

Page: 230: © Lester V. Bergman/Corbis.